Global Challenges in Water Governance

Series editor
Jeremy J. Schmidt
Durham University
Durham, UK

"A beautiful synthesis of the emergence of water governance, its significance in human affairs, and the challenges it entails on a human dominated planet. Short, comprehensive, and easy to read. I can highly recommend it."
—Carl Folke, *Science Director and Co-Founder of*
Stockholm Resilience Centre, Sweden

After a century of massive human interventions into the hydrological cycle, governing water is a critical global concern in the new millennium. Growing evidence that human impacts on the planet are shaping global and local hydrology is challenging long-held assumptions regarding resource management, development, and sustainability. Global Challenges in Water Governance introduces and examines physical, social, and ethical factors that affect how relationships to water amongst humans, social institutions, other species, and Earth systems are governed. Each volume in the series tackles issues of critical importance to water governance— from relationships of science to policy, to water politics and human rights, to ecological concerns—in order to clarify what is at stake and to organize the complex contexts in which decisions are made. Broadly interdisciplinary, the series provides fresh, accessible insights across the sciences, social sciences, and humanities from established academics and talented young scholars. Individual books are ideal for educators, as policy primers for governmental and non-governmental sectors, and for researchers whose work is directly or incidentally connected to water issues.

More information about this series at
http://www.springer.com/series/15055

Jeremy J. Schmidt · Nathanial Matthews

Global Challenges in Water Governance

Environments, Economies, Societies

Jeremy J. Schmidt
Durham University
Durham, UK

Nathanial Matthews
Senior Visiting Fellow
King's College London
London, UK

Global Challenges in Water Governance
ISBN 978-3-319-87094-6   ISBN 978-3-319-61503-5   (eBook)
DOI 10.1007/978-3-319-61503-5

Cover illustration: © Harvey Loake

Printed on acid-free paper

This Palgrave Macmillan imprint is published by Springer Nature
The registered company is Springer International Publishing AG
The registered company address is: Gewerbestrasse 11, 6330 Cham, Switzerland

PREFACE

The *Global Challenges in Water Governance* series has three interlocking aims. First, to provide a resource for understanding multiple aspects of the challenges of water governance in an era when humans are impacting the Earth system—and its water—like never before. Second, to draw together different elements of the natural and social sciences that serve to clarify the different aspects of water governance, such as how propositions of water scarcity and water security connect people and planet. Third, to elucidate the challenges of water governance at multiple scales such that decision makers and educators have further resources in thinking about the possibilities, and limitations, of water governance institutions.

In this first volume of the series, Nathanial Matthews and I track the emergence of global water governance. It is a story that unfolds in the larger context of growing environmental awareness in the late twentieth century, advances in Earth sciences, and international efforts in sustainable development that came to prominence after the Earth Summit in Rio in 1992. These factors coalesced—loosely, but recognizably—in projects of integrated water resources management (IWRM) that rose to prominence throughout the 1990s. The goal of integrated management for water was to organize a holistic approach to the triple-bottom line of sustainable development: environment, economy, and society. In this volume, we examine how this triple-bottom line starts to look different when water governance emerges in the new millennium. Efforts to impose uniform institutions, policies, and practices quickly ran into the rough ground of uneven development, environmental variability, and political contests over how, and for whom, sustainable development was pursued. Yet, sustainable development had considerable inertia, both as a common language and as an organizing framework for many types of knowledge. One result is that treating the triple-bottom can be understood in plural—environments, economies, and societies. Recognizing this plurality helps to reflect the turn from management to governance of natural resources now widely underway within and beyond the water sector.

By exploring how water governance emerges in this broader context, this volume provides a historical orientation and a contemporary analysis of why global challenges in water governance are framed as they are and what possibilities and limitations result. As the first in a series that explores particular issues in greater depth, this volume provides context for understanding the emergence of global water governance and content regarding how its particular challenges are the outcome not only of human impacts on freshwater but also from previous attempts to resolve them. It aims to elucidate these challenges from multiple perspectives, to show why some problems are perennial and to identify those that are novel. With these in mind, it suggests possible paths forward in a time when water concerns occupy a central place in responding to the challenges of global environmental change.

Durham, UK, 2017   Jeremy J. Schmidt

ACKNOWLEDGEMENTS

I am grateful to Nate for his commitment to bridging ideas and practice and for bringing that perspective to this project. The Social Sciences and Humanities Research Council of Canada has generously funded my research on water governance. My wife, Sohini, has been wonderfully supportive during the writing of this book, selflessly sharing her time and insights during the transition from one continent to another.

Jeremy J. Schmidt

I would like to thank Jeremy for asking me to be a part of this interesting project. Special thanks to all my colleagues and friends across the water world and beyond who have helped shape my thinking around governance and the interconnectivity of water. Above all I want to thank my wife, Ay, and my son Caio for their patience and support, especially during all the late-night writing.

Nathanial Matthews

.

CONTENTS

LIST OF FIGURES

CHAPTER 1

Global Water Governance: An Overview

Abstract This chapter distinguishes water management from water governance. It provides an overview of international water management from the UN Conference on Water in Mar del Plata to projects of global water governance that began in earnest in 2000. It emphasizes the important role that programs of integrated water resources management (IWRM) played during the 1990s, the problems and potential of which significantly shaped the challenges taken up by global water governance. Through this historical overview, the chapter defines and explains the specific attention that water governance gives to the social and political structures of decision making. As the result of the significant role of IWRM, existing structures of international water management connected water governance to programs of sustainability that aim to maximize outcomes across the triple-bottom line of environmental, economic, and social well-being. The chapter identifies the liberal compromise of sustainable development and the ways in which liberal notions of political and social order have both compelled and constrained notions of sustainable development.

Keywords Integrated water resources management · Structure
Mar del Plata · Governance · Sustainable development

Life on Earth exists between molecular compounds of water and the global hydrological cycle. This fact is no secret, yet it belies a fundamental tension that sits at the heart of the challenges that are central to this book. Namely, that just as water governs the space in which all kinds of life dwell—from worms to whales—so too have humans come to use water in ways that fundamentally alter this global space. Humans have pushed water around in ways that often dwarf the actions of other species, and which are now starting to alter how water functions as part of the Earth system. Thousands of immense dams, miles of canals, industrial pumps sucking water up from underground, and bustling megacities present just a few of the direct ways in which people are reshaping the planet's water system to suit themselves. Added to these direct diversions and interventions are all the indirect ways humans affect water when they alter landscapes to grow food, harvest forests, produce leather, extract fossil fuels, and the myriad other activities in which effects on water may not appear obvious straightaway, but are nonetheless significant and, often, connected. Together, the cumulative impacts of direct and indirect human water uses have produced global problems that are both exceptional and uneven; never before has there been such an audacious water grab and yet not everybody partakes in, and indeed many suffer from, the colossal interventions humans have made on the global water system. A rapidly changing climate, rising populations, and increased demands for water suggest that this water grab may further intensify.

This book offers a way to understand current and future challenges in global water governance, which has emerged in the twenty-first century as a framework for connecting the knowledge required for effective planning and management to the social and political structures in which knowledge is produced, decisions are taken, and interventions are pursued. The challenges facing global water governance are chock full of both history and happenstance, and so, a critical question to ask is: why governance? That is, how and why did governance become so central to understanding global water challenges, and how does it differ from other strategies, such as those of water planning or management? The question is not simply academic because how water challenges are framed, discussed, and understood has important repercussions on what actions may or may not be taken, who or what is considered to be affected, who is consulted, and much else besides. In this sense, answering the question "Why water governance?" aids our understanding of how responses to water challenges account for, and seek to coordinate, not only local and regional concerns, but also broader global dynamics. These overlapping and interwoven considerations are covered in three related areas that this book takes up: environments, economies, and societies. Each of these is an important area in which water challenges figure centrally, and critically, each of these areas has a longer history of framing understandings of the social and political processes that connect water to broader efforts to achieve sustainability. Although we present these separately, they are in practice deeply connected, where social and economic considerations are tightly woven with environmental contexts.

Understanding global water governance challenges in relation to sustainability's "triple-bottom line" of environment, economy, and society provides a starting point for understanding how and why governance has become central to decision making at multiple scales, from cities and farms to countries and global finance. This triple-bottom line arose as numerous actors competed for influence as sustainable development took shape within the context of post-colonialization, international development, and economic globalization in the latter half of the twentieth century (Macekura 2015). Emerging from the rough-and-tumble of environmental politics, sustainable development was a compromise position between the liberal economic order that gained global hegemony during the last half of the twentieth century and the form of environmentalism that was both compelled and constrained by it (Bernstein 2001). In this sense, sustainable development emerged as a compromise position in debates over the finite resources and capacity of Earth versus the contemporary economic paradigm in which growth is central to social and political order (Sabin 2013). The tentative resolution to these debates—the compromise—was to seek equilibrium among liberal notions of economic growth and social order through development deemed environmentally sustainable. Understanding the categories of environment, economy, and society as though there is only one type of each, however, can be somewhat limiting insofar as this may crowd out the alternatives environmental governance may wish to understand (Bernstein 2002). In other words, there was more than one way to move beyond the impasse between environmental limits and economic growth. Moreover, the particular solution of "sustainable development" has, since its uptake, shaped both environmentalism and governance in ways peculiar to that solution. As such, and to keep this in mind, we emphasize the plural and open nature of environments, economies, and societies. In so doing, we acknowledge the important role of the triple-bottom line in global water governance while recognizing that it has never gone uncontested.

To situate the triple-bottom line, each of environments, economies, and societies has its own chapter dedicated to understanding how multiple perspectives regarding science, politics, knowledge, and cultural practices have affected the ways in which water was brought under the larger sustainability agenda. Before moving to those three areas, this chapter introduces the distinctive elements that shaped the rise of global water governance, which has a history somewhat different than other areas of environmental governance, such as desertification, that also received increasing global attention in the 1970s (see Davis 2016). It begins by distinguishing water governance from water management. This will help to show how, later, water governance reframed water management challenges as it was solidified as a central part of major global initiatives all the way through to achieving the Sustainable Development Goals (SDGs) adopted by the United Nations in 2015. Through this introduction, we also start a conversation that continues throughout the book regarding what makes something a global challenge. Is it because the problem threatens the entire, or a significant part of, the environment? Or is it because the complex dependencies of international trade make water challenges a risk to an increasingly connected global economy? Or is it because the unequal inputs and outputs of human impacts on the water system produce social outcomes both uneven and unjust? Such questions are critical to understanding the benefits and limits of framing global water challenges in terms of governance. In the concluding chapter, we draw attention to the challenges of achieving justice and fairness in addressing global challenges in water governance and, indeed, using governance as the frame for understanding these challenges. This final challenge is especially daunting. Yet, it must be kept central because the global water system is, above all and unavoidably, something that is shared not only across the triple-bottom line of sustainability, but also across cultures, religions, species, and peoples.

FROM MANAGEMENT TO GOVERNANCE

Water governance rose to global prominence after 2000, when an international conference in The Hague solidified its importance in addressing the global water crisis. Prior to then, it was much more common to find water challenges being addressed as issues of water management or planning that were oriented to maximizing the benefits to be derived from water. As it is typically used in resource contexts, management refers to the sites of decision making that affect the environment—from farmers deciding when to irrigate their crops to city managers making decisions on the design of storm runoff systems (Mitchell 2002a). By the time experts and officials gathered in The Hague, and for reasons we discuss below, the emphasis of water management on gaining comprehensive knowledge of how decisions were made and of seeking more rational models for those decisions didn't account well for the politics or limitations of either knowledge or institutions (Brugnach and Ingram 2012). On the other hand, the traditional focus of water management on the actual decisions affecting water required organizing such a cacophony of water use decisions that it was hard to see how any single framework could be applied. So the shift toward governance was more than just one of terminology. Water governance promised to overcome the limitations of previous ways of assembling and addressing water challenges and to open up new forums for seeking solutions. The Ministerial Declaration from The Hague was one of the first international calls for "good governance" and was itself part of a longer process of what the World Water Council called "making water everybody's business"—that is, governance expanded coordination from single management frameworks to include multiple actors, institutions, environments, and agendas that affected, and were affected by, water use decisions (Cosgrove and Rijsberman 2000).

The need to find a more robust coordinating framework for addressing water challenges arose parallel to a broad consensus regarding water scarcity, which was first recognized globally in 1977. Despite several decades of research and awareness raising activities regarding the threats posed by water scarcity, there were still not sufficiently robust responses to this challenge by the turn of the millennium. The result was that water scarcity was becoming acute enough in many places that experts began to explicitly link scarcity to concerns over security; various types of conflicts may arise as water scarcity intensifies, both between directly competing human water uses but also due to effects on other species, and ecological processes more broadly. Water security considerations may therefore arise as the result of acute conflicts over scarce water, or as chronic water scarcity undermines the social or environmental conditions for successful societies. Concerns over the conflicts and conditions affected by scarcity are not limited, however, to lack of water alone. They also include not having enough water of sufficient quality at the right times and places. Making the matter more complicated, having too much water is also a security problem when critical infrastructure, lives, and livelihoods are put at risk by floods. Conflicts can arise here too, particularly given the siting and operation of flood management infrastructure. Given this complexity, a debate arose over how to understand water security (see Chap. 4). Was water security something that can be measured in terms of risk or an issue best approached by integrating uncertainty into decision making? Despite differences over how to answer this question, all sides agreed that improved water governance was essential.

To see how water governance emerged in response to the limitations of managerial approaches to water scarcity and security, it is helpful to provide more context to the specific challenges contemporary calls for governance respond to. As a first cut, water governance can be understood as connecting the knowledge required for effective planning, and the factors affecting actual water management decisions, to the social and political structures in which knowledge is produced and decisions are taken. For much of the twentieth century, however, state water planning and management was heralded as the most effective way to coordinate water uses across a range of social and political processes. Histories of water planning and management of course go back much further, but the large-scale, industrial processes and infrastructure that characterize many contemporary water challenges crystallized relatively recently as projects of modernization required increased water inputs for production (see Wittfogel 1957; Molle et al. 2009). Coordinating water uses on an industrial scale was often accomplished by gathering quantitative data on hydrological availability and variation, and combining it with political will, technology, and both public and private finances to produce desired outcomes. A central aspect of water planning was a belief that by gaining a comprehensive picture of water, decisions could be made in ways that were both objective and rational. In this sense, water management and planning was premised on getting as close to a total vision of water as possible through hydrology and then adjudicating competing demands based on that picture (Scott 2006).

Initially, what counted as objective was widely agreed upon—facts produced by scientific measurements and calculations based on a combination of engineering and hydrology. Ensuring that decisions taken from this objective picture were rational likewise depended on removing bias, such as through techniques of cost-benefit analysis (see Chap. 3).

Justifying water management and planning as both objective and rational was vital because as modern, industrializing states grew, it was readily apparent that not only were demands for water growing but that water infrastructure also needed to be designed for more than a single purpose. State planning needed to maximize the benefits derived from water for multiple purposes. This resulted in a shift away from projects designed for single purposes, such as building a dam so that its reservoir provided irrigation water, toward what became known as multipurpose development, such as when the waters stored by dams are coordinated for use in flood control, hydropower, and irrigation. But what was the most effective use of water across these multiple purposes? Answering this question without bias required an objective account that did not have the appearance of favoring or oppressing any particular group of water users. Further complicating matters was that not all water users had the same kinds of demands. Providing a steady supply of water for cities looks much different than the seasonal demands for irrigation. Soon, then, multipurpose water projects were connected to achieving maximum results through multiple means. Simply increasing water supplies was one option, and remained a large part of many planning responses, but other means were also available. Enhancing efficiency through new technologies, for instance, could be combined with new regulations or pricing mechanisms to prevent waste and enhance productivity. In sum, as demands for water grew in scale and scope during the twentieth century, there was a trend in planning from single-use, single-means projects toward those designed as multipurpose, multiple-means projects (White 1969).

There were also competing projects and visions of modernization itself, especially as the industrializing powers of the Cold War sought to gain global influence in ways that demanded—and often strategically used—ever more water (see Ekbladh 2010; Sneddon 2015; Swyngedouw 2015). By 1977, the pace and scale of industrialization were recognized as requiring international coordination. That year, at the first UN Conference on Water in Mar del Plata, Argentina, domestic programs of water management and planning were scaled up into a program of Rational Planning that became a founding part of the global water agenda (Biswas 1978). But despite the promise of Rational Planning, not everything was as objective or unbiased as initially hoped. On the one hand, large-scale uncertainties about hydrology remained, while the short- and long-term consequences of many projects were not always well known at regional or local scales. In fact, when global leaders met in Mar del Plata, only 2 years had passed since UNESCO concluded its International Hydrological Decade, which was the first international scientific attempt to quantify global hydrology. So, while there was a considerably more accurate map of the planet's water, the global water atlas still had many blurry and even empty portions. Planners, in short, lacked the full, comprehensive picture of water required for Rational Planning; some guesswork remained.

More fundamentally, however, challenges arose regarding just how "rational" water management and planning actually was. Decisions made within institutions and bureaucracies didn't always follow the rules (often for political reasons), while at other times competing bureaucrats had differing interpretations of rules and data (Espeland 1998). But Rational Planning faced two other problems. First, different groups may frame problems differently, so the variables that are important to one group may not be so to others. Second, even if different groups agree on which variables matter, they may disagree on what weight to assign to them or what consequences are likely to follow from different plans (Lindblom 1959, 1999). Given the fact that water use decisions play out in contexts shaped by multiple perspectives, water management and planning increasingly started to look like other "wicked problems." That is, problems where intractable differences regarding the definitions, solutions, or problem frames mean that no resolution is likely to meet a single, uncontested standard of rationality (Rittel and Webber 1973).

In the decade after Mar del Plata, the broader environmental agenda was increasingly framed around the concept of sustainable development, which was given its fullest expressions by the World Commission on Environment and Development (1987) in the volume, *Our Common Future*. When *Our Common Future* was published in 1987, the assessment of it by water professionals was unforgiving. They charged that sustainable development had been articulated without adequate appreciation of water (International Water Resources Association 1991); later, one prominent hydrologist went so far as to call it "water-blind" (Falkenmark 2001, p. 552). In response to this perceived oversight, water professionals were determined not to allow the emerging sustainable development agenda to retain its hydrological myopia and began to strategize for how best to ensure water was clearly on the agenda at the UN Conference on Environment and Development set for Rio de

Janeiro in 1992. As part of preparing for the Earth Summit in Rio, the decision was made to hold a preliminary conference in Dublin to sharpen the message connecting water to sustainable development. The Dublin conference also created an opportunity to address some lingering problems, particularly those associated with being unable to gain a complete quantitative picture of global hydrology and the "wicked problems" that prevented uncontested, rational decisions.

Recall that part of the impetus for Rational Planning was to coordinate multipurpose projects with multiple means. Throughout the 1980s, this goal faced yet another challenge as the result of a patchy, but sustained retreat from the position that states were the right kind of institutions for gaining a comprehensive knowledge of societies or economies. In place of governments, economic and transactional logics were the order of the day not only for resource planning but also for understanding social and political order more broadly (Brown 2015). Although many water professionals called for institutional renewal rather than abandonment, the shifting role of states opened a space that new forms of economic reasoning were often the first to fill (Ingram et al. 1984). It was not as though economics was new to the water game. Cost-benefit analysis, development loans for large dams, and financing arrangements for infrastructure were by then a significant part of the multipurpose development and international politics (Mitchell 2002b). Rather, what was new was the priority assigned to economics to provide information about how water was best valued. As Chap. 3 considers more closely, this had several unanticipated consequences. Yet, as water professionals gathered in Dublin to hammer out what basic principles should be advanced at the Earth Summit in Rio, economics found a new place within the social and environmental concerns over water scarcity in what became known as the four Dublin Principles: [1] "Freshwater is a finite and vulnerable resource, essential to sustain life, development and the environment...[2] Water development and management should be based on a participatory approach, involving users, planners and policy makers at all levels...[3] Women play a central part in the provision, management and safeguarding of water...and [4] Water has an economic value in all its competing uses and should be recognized as an economic good" (Dublin Statement 1992).

With the Dublin Principles in hand, water professionals called for a step away from state-led planning and sought to coordinate multipurpose, multiple-means water projects through the concept of integrated water resources management (IWRM). Since states were no longer seen as having sufficient resources for comprehensive, Rational Planning, integrated water resources management would instead focus on identifying and integrating the many different ways in which water use decisions were actually made. Here, Rational Planning was not being entirely abandoned. In fact, gaining a full picture of human impacts on water remained as important as ever, and states had an important role in providing data and re-regulating the water sector in ways that allowed for the kind of "integration" being pursued through the Dublin Principles. Here, because states would still retain important obligations regarding laws, policies, and rights, the scientific knowledge of hydrology that was used in planning was to be integrated with a broader set of concerns that were not easily tracked by the state but which could be known through closer attention to gender dynamics, public participation, and especially the preferences revealed by economic transactions. After Rio, the concept of IWRM rose rapidly through the ranks of the global water community as new professional networks, publications, and international conferences were organized to promote it as a framework for decision making that aligned with the broader agenda of sustainable development (Conca 2006).

Key to the global uptake of IWRM was the formation of the Global Water Partnership (GWP), which was conceived of in 1995 as a joint initiative of the World Bank and the United Nations Development Programme. After its founding in 1996, the GWP began to articulate the central tenets of IWRM and, in the years that followed, published a series of technical advisory papers for governments and development organizations. At the turn of the millennium, the GWP (2000, p. 22) published one of the most widely cited definitions of IWRM as:

> a process which promotes the coordinated development and management of water, land and related resources, in order to maximize the resultant economic and social welfare in an equitable manner without compromising the sustainability of vital ecosystems.

This definition of IWRM by the Global Water Partnership was released in the same year as the meeting in The Hague, where decision makers were increasingly connecting water challenges to issues of governance. Why was a shift toward governance happening now, at the same moment when years of international water management efforts were finally starting to consolidate in clearer definitions of IWRM? Part of the reason was that, like Rational Planning before it, IWRM was not without its difficulties and deficiencies when it came to how it coordinated water projects. Indeed, by the early 2000s, even former proponents of IWRM had started leveling critiques about its broad and at times vague prescriptions. These were deemed to have failed for many reasons: IWRM was too ambiguous to provide clear inputs to policy, it failed to appreciate existing institutions (i.e., rights to land or water), and when it did manage to coordinate water uses, it often did so using primarily technical criteria that had the effect of marginalizing key water users, notably women and others suffering social or political inequalities (Biswas 2004a; Blomquist and Schlager 2005; Jeffrey and Gearey 2006). Water experts also bemoaned the circuit of international water conferences that gathered to repeat messages regarding the need for improved planning and management that, at least within the water community, were by then widely agreed upon (Biswas 2004b; Gleick and Lane 2005).

A key reason that IWRM wasn't able to deliver on its promise was that the institutions it sought to coordinate hadn't been designed with water at their core. Frequently, state institutions of health, economics, development, and even the environment had been designed and structured in ways that affected and required water, but in which coordinating water uses was not a priority. The result was that IWRM faced an uphill battle not only because of existing institutional constraints but also because it was articulated in predominantly water-centric terms that required major political change. In part, this water-centric approach can be seen as a pendulum swing motivated by efforts of water professionals to get water on the sustainable development agenda after their interpretation of *Our Common Future*. Whatever the mix of factors, IWRM foundered on this ambitious attempt to position water at the center of sustainable development efforts. By 2004, and after becoming a key player in IWRM programs throughout the 1990s, even the World Bank (2004) was backing away from the concept even though it retained its commitment to the Dublin Principles (see also Goldman 2005).

As the twenty-first century got underway, it became evident that efforts in IWRM were often misaligned with the political and social structures through which knowledge is produced and actions are taken. Indeed, experts started to call for a new paradigm in the water sector. These voices grew louder as knowledge of human impacts on the global water system accumulated, and the institutions put in place for planning and management were increasingly stressed by growing water demands (Postel et al. 1996; Gleick 2000). As Chap. 2 shows, experts started to identify how the growing impact of human water uses, in combination with global environmental change, led to an unavoidable conclusion: Major institutional overhauls would be required to sustainably managing water and the scale of change rendered any single management framework inadequate. All was not lost, however. The sharpened definitions of IWRM and the lessons learned from attempts to address the shortcomings of Rational Planning could be identified for where and how they were misaligned with social or political structures that affected decisions. This new focus could arrange the techniques of water management and planning with respect to the gaps, problems, and politics of different contexts of governance.

In the view of the Global Water Partnership, a focus on governance could expand the range of perspectives and stakeholders included in, and affected by, decisions regarding water. By expanding opportunities to coordinate social and political inputs into decisions, governance held the promise of producing results that would be more effective in achieving sustainability. Through effective governance, water development would not be the sole purview of government but rather refer, "to the range of political, social, economic, and administrative systems that are in place to develop and manage water resources, and the delivery of water services, at different levels of society" (Rogers and Hall 2003, p. 7). The result was that both planning and management needed to be repositioned to account for the social and political structures through which decisions were taken, and with respect to the range of different kinds of knowledge used in these decisions. This knowledge ranged from the place-based, local knowledge of communities to the empirical picture of the planet offered by Earth system science.

Like the program of international water management before it, the turn to governance had a global focus. It faced, and continues to face, many of the constraints that dogged the attempt(s) of sustainable development to align environmental policies with a liberal order premised on economic growth (see Brown and Garver 2009). It has also sought to align new approaches to environmental governance with growing recognition that water management, like other resource management sectors, faces a dual pressure to stop human water uses short of planetary boundaries while allocating and distributing environmental goods justly and fairly (Dearing et al. 2014). The compromise of liberal environmentalism, in short, has not gone away. But does increasing attention to social and political structures of governance provide something new? Or does it simply push the problem from one scale (the managerial point of decision making) to another, this time focused on social and political structures? Regardless of our responses to this question, the turn to governance must also find common cause among different approaches to water that vary across time, space, and cultures, regarding what constitutes a good social or political arrangement.

GOVERNANCE: BOTH GLOBAL AND GOOD?

In its simplest definition, the term structure refers to the arrangements and relations between the parts or elements of something more complex. At a molecular level, for instance, combinations of hydrogen and oxygen have many different potential structures. The most commonly referred to in water policy is H_2O, but these molecules can also combine into H_3O^+ (hydronium) or ^2H_2O (deuterium). In similar fashion, social or political structures refer to the arrangements and relations between individuals, classes, or groups that are part of something more complex, such as societies (Bourdieu 1977). For a global economy to run requires different elements, such as currencies, property laws, international trade agreements, and notions of credit and debt (to name only a few elements) to be arranged in a particular way. Similarly, the social and political structures governing water require numerous arrangements in order to succeed: infrastructure for water delivery, finance for operations and maintenance, sound knowledge of hydrological variability, assessments of environmental needs, estimates of risk, principles for water sharing, recognition of rights, and so on. Designing these governance arrangements is not a simple task, especially given the different meanings of water held by individuals and groups (Strang 2004). As global water governance was developed, it sought to avoid the prescriptive, water-centric troubles that plagued IWRM, yet it also needed an approach to social and political structures that could allow for diversity while maintaining (the compromise of) sustainable development.

The strategy of global water governance has been to avoid judgments regarding the desirability of different social or political orders and instead to argue that there are more basic criteria that are desirable as a shared basis for global coordination. Accordingly, "good water governance" is governance based on shared commitment to these criteria—while the criteria themselves can be expressed through the customary practices of social or political structures that may differ considerably. Numerous definitions of good governance have been offered (see Box 1), with many combining criteria such as enhanced transparency in decision making, fair processes for participation, and subsidiarity—the devolution of decision-making powers to the most appropriate level. These criteria are frequently touted as principles even though they are generally declared to be inherently good (i.e., treated as criteria) rather than being rationally defended as right (i.e., defended as sound principles). For example, "transparency" is often touted as a principle of good governance since it offers a bulwark against corruption. But merely being transparent does not automatically entail good consequences. No matter how transparent governance is, certain actions are still not right, such as discriminating against particular classes, races, or genders. So "transparency" is not an unqualified good. As Chap. 4 details, the slippage between the right and the good would later lead governance practitioners to focus more closely on considerations of ethics and fairness in the processes used to reach decisions and to monitor results. A principle reason for focusing on ethics and fairness was the assumption that if the procedures for reaching decisions were seen as fair by those involved in governance exercises, then the outcomes were more likely to be accepted as legitimate. That is, the problem of trying to get all water problems to fit into a single management model like IWRM could be solved by developing a shared basis for reaching decisions that did not care what the end "model" looked like so long as it was effective in arranging environmental, economic, and social considerations. Rather than trying to solve wicked problems through rational agreement, or using state decision-making power to adjudicate dilemmas, governance would instead focus on normative means for generating social and political legitimacy (cf. Bernstein 2002).

> **Box 1: Definitions of good water governance**
> **Good Governance**
> • "Good governance is epitomized by predictable, open, and enlightened policy making (that is, transparent processes); a bureaucracy imbued with a professional ethos; an executive arm of government accountable for its actions; and a strong civil society participating in public affairs; and all behaving under the rule of law." ~ World Bank (1994, vii).

- "Based on over a decade of experience with development progress and challenges...effective governance institutions and systems that are responsive to public needs deliver essential services and promote inclusive growth, while inclusive political processes ensure that citizens can hold public officials to account. In addition, good governance promotes freedom from violence, fear and crime, and peaceful and secure societies that provide the stability needed for development investments to be sustained. Women are crucial partners in all these processes." ~ United Nations Development Programme (UNDP 2014, p. 4).
- "The [OECD] Principles [on Water Governance] are rooted in broader principles of good governance: legitimacy, transparency, accountability, human rights, rule of law and inclusiveness. As such, they consider water governance as a *means* to and *end* rather than an end in itself, i.e the range of political, institutional and administrative rules, practices and processes (formal and informal) through which decisions are taken and implemented, stakeholders can articulate their interests and have their concerns considered, and decision-makers are held accountable for water management." ~ Organisation for Economic Co-operation and Development (OECD 2015, p. 5).

The challenge of ensuring that governance was effective in a way that achieved a measure of fairness was key to achieving good water governance. But achieving good water governance did not start from a blank slate. Governance efforts had to contend with many different types of communities, actors, and institutions, many of which had long-standing grievances that required other sorts of remedies, such as those provided for by courts. In these cases, governance was constrained in ways not always widely recognized, such as when processes themselves appear fair from one perspective but nevertheless contribute to injustice (Schmidt and Peppard 2014; Matthews and Schmidt 2014). For instance, in many countries, the rule of law imported initially by colonizers continues to structurally marginalize the claims of indigenous peoples. Thus, appealing to the "rule of law" as a criterion for good governance can build inequality into the structure of governance itself by arranging individuals or groups into unequal relationships. An additional problem that we consider in this book is the fact that as the collective human impact on the

Earth's water systems grew, many of the factors affecting water governance were well beyond the scale of any particular level of decision making. Good water governance, then, also needed to be global.

At the same time that water governance was gaining international prominence, numerous studies painted a picture of human interactions with the planet characterized by both rapid environmental change and growing insecurity for both humans and ecosystems (Meybeck 2003; Vörösmarty et al. 2004). These human impacts were of such a magnitude that they were not simply pushing water around within a relatively stable hydrological system, but actually changing the functioning of the Earth system itself (Milly et al. 2008). As we discuss further in the next chapter, anthropogenic forcing on the global water system added to the urgency with which global governance was pursued because it undermined the assumption of water planning and management assumptions regarding environmental stability. Assumptions that water's variability fluctuated within natural limits were historically a common element used in assessing the probability of high-magnitude, low-frequency events, such as extreme floods or droughts. With anthropogenic climate change pushing the Earth system into a "no analogue" situation unlike any that people have previously encountered, however, water managers increasingly argued that assumptions of stable hydrological variability were no longer tenable (see Steffen et al. 2004).

Growing recognition of human impacts on the global water system soon collided with the rejection of the idea that there was any single institutional format for governance that could be either centrally coordinated or uniformly applied to all contexts. In short, there were no institutional panaceas (Ostrom 2007; Meinzen-Dick 2007). Rather, what was needed was a shift in perspective, from a view where water sat at the center of a broad effort to coordinate decisions, such as IWRM had sought, to a view where it was recognized that humans were already deeply integrated with the global water system—just not on the terms hoped for by proponents of sustainable development (Schmidt 2017).

The search for governance that is both global and good remains a key challenge today. In the remaining chapters of this book, we examine how these two considerations are taken up in the range of considerations that governance seeks to address across environmental, economic, and social domains. For each area, we consider how the production of knowledge at different scales—from the local to the global—affects and shapes what counts for or against good water governance. Different ways of knowing water often result in ambiguity, conflict, and even stalemates, but they are central to efforts to reach decisions that avoid the pitfalls of relying exclusively on water planning or management (Brugnach and Ingram 2012). Our analysis follows others who have pointed out that the ways in which knowledge is produced cannot be separated in a neat and tidy way from how we choose to live. Rather, the production of social and scientific knowledge is closely connected to considerations of order and governance (Jasanoff 2004). In the case of water, the stakes are high. The challenges of seeking water governance formats that are both global and good must be thought about with a view toward not only ensuring just outcomes but also achieving those ends through means that are legitimate and fair—not simply "effective." This is especially the case when our knowledge is partial, when existing social and political structures are unequal, and when decisions must be made in an era of accelerating global environmental change.

REFERENCES

Bernstein, Steven. 2001. *The Compromise of Liberal Environmentalism*. New York: Columbia University Press.

Bernstein, Steven. 2002. Liberal Environmentalism and Global Environmental Governance. *Global Environmental Politics* 2 (3): 1–16.

Biswas, Asit K. (ed.). 1978. *United Nations Water Conference: Summary and Main Documents*. Oxford: Pergamon Press.

Biswas, Asit K. 2004a. Integrated Water Resources Management: A Reassessment. *Water International* 29: 248–256.

Biswas, Asit K. 2004b. From Mar Del Plata to Kyoto: A Review of Global Water Policy Dialogues. *Global Environmental Change* 14: 81–88.

Blomquist, W., and E. Schlager. 2005. Political Pitfalls of Integrated Watershed Management. *Society and Natural Resources* 18 (2): 101–117.

Bourdieu, Pierre. 1977. *Outline of a Theory of Practice*. Cambridge: Cambridge University Press.

Brown, Peter G., and Geoffrey Garver. 2009. *Right Relationship: Building a Whole Earth Economy*. San Francisco: Berrett-Koehlers.

Brown, Wendy. 2015. *Undoing the Demos: Neoliberalism's Stealth Revolution*. New York: Zone Books.

Brugnach, M., and H. Ingram. 2012. Ambiguity: The Challenge of Knowing and Deciding Together. *Environmental Science & Policy* 15: 60–71.

Conca, Ken. 2006. *Governing Water: Contentious Transnational Politics and Global Institution Building*. Cambridge: MIT Press.

Cosgrove, William J., and Frank R. Rijsberman. 2000. *World Water Vision: Making Water Everybody's Business*. London: Earthscan.

Davis, Diana K. 2016. *The Arid Lands: History, Power, Knowledge*. Cambridge: MIT Press.

Dearing, John, Rong Wang, Ke Zhang, James Dyke, Helmut Haberl, Md Sarwar, Peter Langdon Hossain, Timothy M. Lenton, Kate Raworth, Sally Brown, and Jacob Carstensen. 2014. Safe and Just Operating Spaces for Regional Social-Ecological Systems. *Global Environmental Change* 28: 227–238.

Dublin Statement. 1992. The Dublin Statement on Water and Sustainable Development. Retrieved from: http://www.un-documents.net/h2o-dub.htm. Accessed on 27 June 2017.

Ekbladh, David. 2010. *The Great American Mission: Modernization and the Construction of an American World Order*. Princeton: Princeton University Press.

Espeland, Wendy N. 1998. *The Struggle for Water: Politics, Rationality, and Identity in the American Southwest*. Chicago: University of Chicago Press.

Falkenmark, Malin. 2001. The Greatest Water Problem: The Inability to Link Environmental Security, Water Security and Food Security. *International Journal of Water Resources Development* 17 (4): 539–554.

Gleick, Peter H. 2000. The Changing Water Paradigm: A Look At Twenty-First Century Water Resources Development. *Water International* 25 (1): 127–138.

Gleick, Peter H., and John Lane. 2005. Large International Water Meetings: Time for a Reappraisal. *Water International* 30 (3): 410–414.

Global Water Partnership (GWP). 2000. Integrated water resources management. *Technical Advisory Committee Background Papers No. 4*. Stockholm: Global Water Partnership.

Goldman, Michael. 2005. *Imperial Nature: The World Bank and Struggles for Justice in the Age of Globalization*. New Haven: Yale University Press.

Ingram, Helen, Dean Mann, Gary Weatherford, and Hanna Cortner. 1984. Guidelines for Improved Institutional Analysis in Water Resources Planning. *Water Resources Research* 20 (3): 323–334.

International Water Resources Association. 1991. Sustainable Development and Water: Statement on the WCED Report *Our Common Future*. In *Water: The International Crisis*, ed. Robin Clarke, 182–185. London: Earthscan.

Jasanoff, Sheila. 2004. *States of Knowledge: The Co-Production of Science and Social Order*. New York: Routledge.

Jeffrey, Paul, and Mary Gearey. 2006. Integrated Water Resources Management: Lost on the Road From Ambition to Realisation? *Water, Science & Technology* 53 (1): 1–8.

Lindblom, Charles E. 1959. The Science of "muddling Through". *Public Administration Review* 19 (2): 79–88.

Lindblom, Charles E. 1999. A Century of Planning. In *Planning Sustainability*, ed. K.M. Meadowcroft, 39–65. New York: Routledge.

Macekura, Stephen J. 2015. *Of Limits and Growth: The Rise of Global Sustainable Development in the Twentieth Century.* Cambridge: Cambridge University Press.

Matthews, Nathanial, and Jeremy J. Schmidt. 2014. False Promises: The Contours, Contexts and Contestation of Good Water Governance in Lao PDR and Alberta, Canada. *International Journal of Water Governance* 2 (2/3): 21–40.

Meinzen-Dick, Ruth. 2007. Beyond Panaceas in Water Institutions. *Proceedings of the National Academy of Sciences* 104 (39): 15200–15205.

Meybeck, Michel. 2003. Global Analysis of River Systems: From Earth System Controls to Anthropocene Syndromes. *Philosophical Transactions of the Royal Society B* 358: 1935–1955.

Milly, P.C.D., J. Betancourt, M. Falkenmark, R.M. Hirsch, Z.W. Kundzewicz, D.P. Lettenmaier, and R.J. Stouffer. 2008. Stationarity is Dead: Whither Water Management? *Science* 319: 573–574.

Mitchell, Bruce. 2002a. *Resource and Environmental Management.* Essex: Pearson Education Limited.

Mitchell, Timothy. 2002b. *Rule of Experts: Egypt, Techno-Politics, Modernity.* Berkeley: University of California Press.

Molle, Francois, P. Mollinga, and P. Wester. 2009. Hydraulic Bureaucracies and the Hydraulic Mission: Flows of Water, Flows of Power. *Water Alternatives* 2 (3): 328–349.

Organisation for Economic Co-operation and Development (OECD). 2015. *OECD Principles for Water Governance* [Welcomed by Ministers at the OECD Ministerial Council Meeting on 4 June 2015). Paris: OECD.

Ostrom, Elinor. 2007. A Diagnostic Approach for Going Beyond Panaceas. *Proceedings of the National Academy of Sciences* 104 (39): 15181–15187.

Postel, Sandra, Gretchen Daily, and Paul Ehrlich. 1996. Human Appropriation of Renewable Fresh Water. *Science* 271: 785–788.

Rittel, Horst W.J., and Melvin M. Webber. 1973. Dilemmas in a General Theory of Planning. *Policy Sciences* 4: 155–160.

Rogers, Peter, and Alan W. Hall. 2003. Effective Water Governance. TAC Background Papers No. 7, Evander Novum, Sweden.

Sabin, Paul. 2013. *The Bet: Paul Ehrlich, Julian Simon, and Our Gamble Over Earth's Future.* New Haven: Yale University Press.

Schmidt, Jeremy J. 2017. *Water: Abundance, Scarcity, and Security in the Age of Humanity.* New York: New York University Press.

Schmidt, Jeremy J., and Christiana Z. Peppard. 2014. Water Ethics on a Human Dominated Planet: Rationality, Context and Values in Global Governance. *Wiley Interdisciplinary Reviews: Water* 1 (6): 533–547.

Scott, James C. 2006. High Modernist Social Engineering: The Case of the Tennessee Valley Authority. In *Experiencing the State*, ed. L.I. Rudolph and J.K. Jacobsen, 3–52. Oxford: Oxford University Press.

Sneddon, Christopher. 2015. *Concrete Revolution: Large Dams, Cold War Geopolitics, and the U.S. Bureau of Reclamation*. Chicago: University of Chicago Press.

Steffen, W., A. Sanderson, P.D. Tyson, J. Jäger, P. Matson, B. Moore III, F. Oldfield, K. Richardson, J. Schellnhuber, B. Turner II, and R. Wasson. 2004. *Global Change and the Earth System: A Planet Under Pressure*. Berlin: Springer.

Strang, Veronica. 2004. *The Meaning of Water*. New York: Berg.

Swyngedouw, Erik. 2015. *Liquid Power: Contested Hydro-Modernities in Twentieth-Century Spain*. Cambridge: MIT Press.

Vörösmarty, Charles J., D. Lettenmaier, C. Lévqêue, M. Meybeck, C. Pahl-Wostl, J. Alcamo, W. Cosgrove, H. Grassl, H. Hoff, P. Kabat, F. Lansigan, R. Lawford, and R. Naiman. 2004. Humans Transforming the Global Water System. *EOS* 85: 513–516.

United Nations Development Programme (UNDP). 2014. *Governance for Sustainable Development: Integrating Governance in the Post-205 Development Framework*. New York: UNDP.

White, Gilbert F. 1969. *Strategies of American Water Management*. Ann Arbor: University of Michigan Press.

Wittfogel, Karl A. 1957. *Oriental Despotism: A Comparative Study of Total Power*. New Haven: Yale University Press.

World Bank. 1994. *Governance: the World Bank's experience*. Washington D.C.: World Bank.

World Bank. 2004. *Water Resources Sector Strategy: Strategic Directions for World Bank Engagement*. Washington, DC: World Bank.

World Commission on Environment and Development. 1987. *Our Common Future*. Oxford: Oxford University Press.

CHAPTER 2

Environments

Abstract This chapter examines how environmental concerns have been addressed in global water governance. It begins with a synopsis of some of the persistent and emergent factors affecting the health of freshwater ecosystems. It then situates the emergence of global water governance in the context of environmental values that gained political salience during the 1960s onward and which structured initial links of science to policy. This chapter shows how environmental concerns led to a commitment to developing more routine world water assessments. These have evolved over time and in response to the growing knowledge of the planet provided by the Earth sciences. Higher resolution accounts of both surface and groundwater availability are critical for global water governance, but can also belie the concrete environmental complexities at regional or local scales. To examine this complexity, the chapter concludes with a case study of the Mekong River Basin to show how western agendas of global water governance and IWRM have clashed with local and regional dynamics.

Keywords Ecosystems · Groundwater · Earth science · Earth system Values · Environmentalism · Mekong River Basin

Global water governance emerged parallel to, and was influenced by, broader environmental movements in the latter half of the twentieth century. As a consequence, the concerns of global water governance have

© The Author(s) 2017   21
J.J. Schmidt and N. Matthews, *Global Challenges in Water Governance*,
Global Challenges in Water Governance, DOI 10.1007/978-3-319-61503-5_2

been influenced by the western ideas and values that have dominated environmentalism. This has led to contests over how to pursue global water governance because western ideas and values toward the environment have significant blind spots regarding class, race, and gender that affect how others, particularly the poor, are perceived with respect to environmental goods (see Guha 2000). These contests have conditioned sustainable development since the 1980s (Lélé 1991). In fact, drawing diverse views of the environment into a single framework for governance was the ostensible goal of sustainable development and the stated view in *Our Common Future* that: "The Earth is one but the world is not" (World Commission on Environment and Development 1987). So, even though there has always been consensus that water's vital role in providing the environmental conditions for all life must be given utmost priority, it has nevertheless remained challenging to develop a structure for governance that can accommodate multiple social perspectives toward a shared planet.

This chapter begins by providing a snapshot of the importance of freshwater ecosystems. This importance cannot be overstated and, precisely because so much is at stake with respect to water, is the site of numerous governance dilemmas and challenges. This chapter then situates the emergence of global water governance in the context of broader environmental concerns that gained political salience from the 1960s onward. This historical element is critical because, as Chap. 1 noted, water professionals pushed ideas of IWRM with greater energy when they perceived that sustainable development did not give water the attention they thought it warranted. It is also important because broader cultural perspectives toward environmental concerns influenced the values and judgments used to link science to policy. In global water governance, this link was forged initially by the American Gilbert White (1978), who provided the first global assessment of water resources and needs in preparation for the UN Conference on Water in Mar del Plata. From this beginning, and as the chapter considers, world water assessments have evolved over time and in response to the growing knowledge of the planet provided by the Earth sciences. These provide higher resolution accounts of both surface and groundwater availability and the effects of humans on planetary systems. Today, these assessments are critical for global water governance, but they can also belie environmental complexities at regional or local scales. The result is that global water governance must navigate a constant tension between multiple kinds of environments—from arctic to arid—that water supports and the complex ways these different environments affect the global water system. To show how the connection between local contexts and global systems is shot through with governance challenges, the chapter concludes by examining the Mekong River Basin as a case study in how a western-led agenda of global water governance and IWRM has clashed with local and regional dynamics.

WATER'S ENVIRONMENTAL IMPORTANCE

Water's links to the environment and to ecosystem services have fluctuated historically and in relation to different cultural demands. For hydrologists, it is critical to think about these dynamics not as distinct spheres of humans and nature but rather as a set of mutual, coevolving relationships (Falkenmark and Rockström 2004). These relationships are complex and multifaceted. They are also characterized by change, which makes understanding the environment a moving target (Folke 2003). Today, as thousands of dams, millions of cubic meters in reservoir storage, rapidly expanding megacities, and kilometers of pipes and canals wring the planet to quench human demands, the global water system is moving in novel ways as the cumulative impacts of humanity push the Earth system beyond the bounds of natural variability (Lehner et al. 2011; McDonald et al. 2014; Rockström et al. 2014). A significant driver of these changes is anthropogenic climate change, which is altering the Earth system as a whole with significant effects for the global water cycle and for assumptions about natural variability that have historically provided water management with a set of parameters for planning (Milly et al. 2008). Given the massive impacts of human water uses, it is now fair to say that water systems are not only moving—they are also morphing into new configurations as human activity alters the Earth system (Schmidt 2017b).

Earth's freshwater ecosystems are home to approximately 126,000 species, yet they are some of the most heavily altered and degraded ecosystems on the planet; these ecological problems, in part, issue from poor governance and lack of coordinated approaches to ecosystem management (Carpenter et al. 2011; UNESCO 2006). Globally, wetlands have declined an estimated 64–71% in the twentieth century alone with degradation continuing mainly due to urban and industrial sprawl and expanding agricultural lands (Gardner et al. 2015). Freshwater species have decreased by 50% since 1970, and rivers are often deprived of the flows necessary for ecosystem functioning or have water quality that is heavily degraded from poorly managed infrastructure or unregulated pollution (Gleick and Palaniappan 2010). These declines are significant not only because changing species configurations affect how ecosystems function but also because these challenges are compounded by patchy governance regimes trying to both conserve freshwater ecosystems and secure the livelihoods of millions of people (Vörösmarty et al. 2010; Russi et al. 2013).

Unfortunately, the vital role of water in biodiversity is not the only environmental challenge for global water governance. Unsustainable groundwater withdrawals for irrigated agriculture are closely tied to the structure of global food supply chains and trade (Dalin et al. 2017). The water demands of rapidly developing economies now push groundwater systems beyond sustainable limits: China's heavy reliance on irrigated agriculture has caused groundwater abstraction to increase from 10 km^3 per year in the 1950s to more than 100 km^3 per year by the 2000s, while overextraction of groundwater in the Central Ganga Plain in India is having significant social and environmental consequences (Wang et al. 2010; Ahmed et al. 2014). The governance dimensions of groundwater are made all the more complicated because many institutions for managing groundwater were initially designed for surface water systems that are ill-suited for the unique challenges presented by the massive expansion of groundwater extraction currently underway around the globe (Birkenholtz 2015). Given that nearly half of the world's seven billion human inhabitants rely on groundwater for drinking water, and that groundwater is critical to the environmental and surface water flows, it is critical that it be governed with much greater care (Alley and Alley 2017).

Climate change presents a major challenge to aquatic environments as precipitation regimes shift, glaciers retreat, and extreme events of flood and drought increase in likelihood and intensity. These effects are compounded by governance challenges, such as efforts to address climate change by touting hydroelectricity as a clean energy alternative despite the fact that reservoirs collectively produce the equivalent of one gigaton of CO_2 emissions annually (Deemer et al. 2016). This amounts to just over 1% of global greenhouse gas emissions, which is small but also not zero. It also does not include the effects of dams on the global carbon cycle (Maavara et al. 2017). Nor does it include emissions or impacts of pouring thousands of tonnes of concrete into infrastructure projects or the large social and environmental impacts of dams. In Brazil, for instance, many of the social and environmental impacts of dams, such as displaced peoples and biodiversity loss, are systematically underestimated (Fearnside 2016). Yet, the push for hydropower in response to climate change is often in calculated as relative to higher polluting fossil fuels that have a large water impact in addition to their greenhouse gas emissions. Conventional and unconventional fossil fuel production requires water inputs that are often orders of magnitude greater than the millions of barrels of oil produced each day. In many cases, the challenges of biodiversity, climate change, and dams are layered upon each other. For example, the Colorado River, a key river in the western USA, suffers from invasive species, point source pollution (e.g., from cities), nonpoint agricultural pollutants, and overextraction (Kennedy et al. 2013; Jones-Lepp et al. 2012). This remains still a partial list: The fate of pharmaceutics and endocrine disrupting chemicals in freshwater systems can also alter aquatic species and ecosystems while raising serious health considerations for human populations (Snyder et al. 2004).

Recognizing the interdependency of humans and ecosystems and developing better governance mechanisms is essential to achieving sustainable management of the planet's freshwater (Matthews 2016; Sedlak 2014). It is also an empirical and ethical imperative for the millions of people who remain without access to reliable and safe drinking water and those who also lack sufficient sanitation (Feldman 2012). Problems arise, however, when it comes to the task of recognizing this interdependency in a way that draws together the multiple social systems currently in place to govern water into a global schema. Indeed, this was and remains a central challenge of global water governance—and the broader agenda of sustainable development to which it often appeals for normative force. One aspect of these complex challenges that is often overlooked is the role of predominantly western values and judgments that have suffused how interdependence should be understood and acted upon to link science to policy.

THE WESTERN ROOTS OF GLOBAL WATER GOVERNANCE

Refusing analytical distinctions between humans and nature is often touted as key to understanding relationships among water, humans, and non-humans as well as to thinking critically about the cultural assumptions that link empirical assessments to normative claims about how governance should proceed (Postel and Richter 2003; Falkenmark and Folke

2010). Despite this, western-led discourses that installed the nature/ society distinction initially are often quickly dissociated from the similarly western-led discourses on water governance seeking to connect science to policy. As many studies have shown, cultural judgments linking empirical assessments of water to social and political institutions cannot be avoided (e.g., Feldman 1995; Espeland 1998). In part, this is because complex systems must always be simplified for the purposes of governance and management, and the judgments made to convert complex systems into manageable units are influenced by historical obligations, cultural values, and interpretations of environmental change. Rather than ignoring or skirting around these cultural influences, a more promising route is to make the implicit cultural roots of global water governance an explicit matter of discussion and debate.

There is no obvious starting point for discussions of values that have deep cultural roots. At the global level, however, key values related to water were articulated by the first director of UNESCO, Julian Huxley (1935, 1943), who swooned over the model of water management that the USA had developed in the Tennessee Valley Authority. Huxley believed the combination of social and environmental engineering undertaken by the TVA could serve as a model for UNESCO programs that would similarly pursue forms of multipurpose, comprehensive resource development and management. Huxley not only liked the combination of natural resource management and social development, he also liked that both were put in service to a liberal approach to social planning. For his part, Huxley (1946) advocated a "scientific humanism" that would vouchsafe social evolution from the clutches of metaphysical and nationalist philosophies. Of course, liberalism was not neutral either. As many scholars have shown, it was common in the postwar era and throughout the Cold War to use water resource development as a tool against communism (e.g., Ekbladh 2010; Sneddon 2015). Critically, however, Huxley was also an avowed eugenicist who believed that social progress could be objectively determined across races and ethnicities. This kind of racism—couching cultural values in the language of science—was not unique to Huxley and also inflected approaches to international development in which water often figured centrally in claims about progress and in legitimizing interventions in the Global South (Escobar 2008, 2012).

The environmental movement that emerged in the 1960s sought to challenge western biases regarding the natural world. Lynn White Jr.

(1967) famously argued that the "roots" of the ecological crisis were to be found in the religious axioms of the West and the unique status assigned to humans. These failed, in White's view, to provide an empirical basis for the world's "emerging, entirely novel, democratic culture." For White, what was novel was not democracy per se but rather the new, global scale it sought to operate at. Indeed, many environmentalists pointed out that scale was a key issue for ecological concerns. Extensive human intervention into ecological systems was eloquently, if devastatingly captured in books like Aldo Leopold's *Sand County Almanac* and Rachel Carson's *Silent Spring*, both of which galvanized environmentalists through their attunement to both science and values. More apocalyptic bells were sounded in Paul Ehrlich's *The Population Bomb* and the Club of Rome's *Limits to Growth*, which warned that the novel rates of industrial consumption and population growth could not be supported by a finite planet. In response, and using techniques that were developed in part to manage water development projects, economists and planners began to focus on growth *rates* instead of on absolute measures of resource limits (Mitchell 2014). In hindsight, even this proved unmanageable; the post-1945 era of global industrialism has been described as the "Great Acceleration," owing to Earth's material resources and energy being channeled into human service at an ever-increasing rate (McNeill and Engelke 2016).

In 1964, and in response to US concerns about the lack of understanding of global hydrology, UNESCO inaugurated the International Hydrological Decade (for an overview, see Schmidt 2017a). The goal of the decade was to produce an account of "water and man"—that is, a universal and objective account of humanity's relationship with water that was based on a scientific worldview (Nace 1969). The decade was the first international scientific collaboration of hydrologists and got underway just as the environmental movement gained global momentum. In 1972, the UN held a conference in Stockholm on the Human Environment, the first global conference of its kind. It was a landmark event and supported by many international agreements, such as the Great Lakes Water Quality Agreement signed between Canada and the USA, as well as numerous new environmental agencies and laws (see Lazarus 2004). Increased awareness and political will on environmental issues was not limited to government agencies. New fields of environmental ethics and economics were developed to cultivate new values—new axioms for the emerging democratic culture that would need to confront growing demands for water and other natural resources. At the first Earth Day in 1970, Gaylord Nelson proclaimed that: "The economy is a wholly owned subsidiary of the environment, not the other way around."

By the time IHD ended in 1974, environmentalism was in full swing. And, when the outcome of the IHD was to convene a global, UN-backed water conference in Mar del Plata, these entanglements were deepened as different governmental and non-governmental actors sought to have their voices heard and views toward water represented (Macekura 2015). They were not always successful, particularly because the aim of global water governance was, at that time, primarily interested in producing a global picture of both humans and water—a scale not particularly conducive to giving serious or sustained attention to cultural differences. As a result, it must be kept in mind that while it was widely agreed to that water is of utmost importance, just *how* that importance was to be understood, interpreted, and governed has always been a matter of contest over which judgments best explain the complex interrelationships among water, humans, and all other life. The focus on new values for linking humans and nature, however, often ignored issues of race, gender, and class in attempts to mobilize environment concerns (Zimring 2016; Gaard 2001). When arguments over new values fell on deaf ears, environmental justice movements took political action against environmental inequalities (Schlosberg 2004, 2010). For feminist scholars, the focus on western values remained a central problem because the putatively "rational" axioms of western thought reified distinctions that were not empirically tenable—notably the unique status of humans as qualitatively distinct from, and superior to, nature (Merchant 1980; Plumwood 2002).

Growing recognition of ecological interdependence intersected with the emergence of global environmental institution building in both the IHD and Mar del Plata. Since that time, contests have continued over how western ideas separating nature from society should be understood, revised, or abandoned. Frequently, scholars have argued that the distinction should be rejected in favor of new concepts, such as the "hydrosocial cycle," that seek to eliminate divisions of hydrology (i.e., nature) from humanity (i.e., society). Some, such as Linton (2010, p. 1), argue that because hydrological science is a social practice—the practice of doing science—that understandings of water are socially constructed to such an extent that "water is what we make of it." On

Linton's account, even the global water crisis is simply a crisis of one way of thinking about and knowing water. It is hard, however, to see how the global crisis of chronically water-deprived children dying from thirst, malaria, or diarrhea is a social construction. So, even if we agree that science is a social practice, it does not lead to the conclusion that anything goes when it comes to water (Schmidt 2014). A more promising route is to consider how numerous social practices produce reliable empirical knowledge (Harding 2015). This also helps to avoid the unethical stance where the knowledge production practices of other cultures are dismissed in a view of "science for West, myth for the rest" (Scott 1996).

Water problems are real—all too frequently they are matters of survival for the world's poorest populations and for many aquatic species. The connections among western values and scientific assessments have positioned global water governance within the broader liberal compromise of sustainable development. As Chap. 4 examines in more detail, the assumptions about "society" that underlie global governance warrant as much or more scrutiny than those regarding "nature." What we are concerned with here, however, is how scientific assessments were used to connect environmental concerns to global water governance and how these connections shifted over time as empirical assessments improved understandings of the Earth system. Forging the links between science and policy involves numerous normative judgments, and forging global agreements is no easy task. Given this, an obvious question arises: How have the subtle, yet shared values that provide legitimacy for the complex judgments involved in governance evolved over time to incorporate the improving empirical picture of Earth's freshwater systems?

Linking "Environment" to Global Water Governance

The IHD provided the first global atlas of the world's water balance (Korzoun et al. 1978). It also proved a decisive and important step marking the "coming of age" of global hydrology (Nace 1980). The knowledge produced by global hydrologists was critical in the project of international water management and the attempt in Mar del Plata to ensure that the holistic and integrated nature of water became a permanent fixture in the public conscious. It was a project that was both scientific and normative, as this statement by the Secretary General to the Mar del Plata conference, Yahia Abdel Mageed, reveals:

It is hoped the water conference would mark a new era in the history of water development in the world and it would engender a new spirit of dedication to the betterment of all peoples; a new sense of awareness of the urgency and importance of water problems; a new climate for better appreciation of these problems; higher levels of flows of funds through the channels of international assistance to the course of development; and in general a firmer commitment on the parts of all concerned to establish a real breakthrough so our planet will be a better place to live in. (quoted in Biswas and Tortajada 2009, p. 5)

At the Mar del Plata conference, delegates sought a rational basis for holistic planning that would use global hydrology as the basis for determining national-level laws and policies (Biswas 1978). It was a challenge shaped by the consequences of accelerating industrial demands of human societies, advances in technology, and of a global population that reached approximately 4.2 billion people in 1977. The main output from the conference was the Mar del Plata Action plan, which in retrospect provided the first international basis for IWRM (Biswas 2004).

As it was conceived at Mar del Plata, integrated water management was a holistic approach that recognized the importance of effective governance. It also recognized deep social challenges, with the 1980s being declared a decade for International Drinking Water and Sanitation by the United Nations. Despite the momentum achieved at Mar del Plata, neither IWRM nor water issues in general garnered the same kind of attention throughout the 1980s that was achieved by sustainable development. Although scientists and academics such as Holling (1986) and Vogel (1997) continued to emphasize the importance linkages between ecosystem health and water governance, it was not significantly addressed at global conferences. Rather, what increasingly occupied water managers was water's relationship with multiple different sectors that presented risks to water. A key conference in this regard was held in 1987 on the topic of human transformation of the Earth. There, Mark L'vovich and Gilbert White (1990) made the first historical estimates of the accelerating use of water by industrial societies over the past three centuries (see also Schmidt 2017a). Water was explicitly linked to global circulation models of the climate by the late 1980s and to emerging concerns over global warming (Gleick 1989). In this sense, risks to water were increasingly being connected to the picture of the planet emerging from the Earth system sciences.

In 1993, Gleick (1993) provided an assessment of the state of the world's water resources that anticipated a biennial assessment process that by 2014 was in its eighth volume. These more routinized assessments were key to developing a global consensus among governance practitioners on the key stressors and risks to the global water system. At the mid-point of the 1990s, water policy expert Sandra Postel, together with key figures in the environmental movement Paul Ehrlich and Gretchen Daily, estimated that humans were appropriating roughly half of the Earth's available supply of renewable freshwater (Postel et al. 1996). Building on these findings, and on growing calls for a world water assessment, the Sixth Session of the Commission on Sustainable Development stated in 1998 that there was a need for regular, global assessments on the status of freshwater resources. As a result, the United Nations World Water Assessment Programme (WWAP) was founded in 2000 with a primary aim to monitor, assess, and report on the world's freshwater resources and ecosystems, water use and management, and to identify critical issues and problems. The main output of WWAP is the World Water Development Report (WWDR). The WWDR reports have developed along with technological and scientific advances in Earth system science to offer increasingly higher resolution accounts of both surface and groundwater availability and the impacts of humans on the planetary system.

Improved assessments of the world's water were paced by the growing prominence of water governance on the global stage. In 1996, with backing from the World Bank and the UN, the Global Water Partnership (GWP) was established as an umbrella organization for promoting and coordinating IWRM and its linkages to water issues on a global scale. The upshot was a period of contest as the management focus of IWRM increasingly rubbed up against emerging concerns of global governance. For some, a key concern was that governance might be relegated as a tool to achieve the objectives of IWRM rather than as an approach with the capacity to deal with the interlinked dynamics of an increasingly globalized world economy (Rogers and Hall 2003; Lautze et al. 2011). It was an important consideration as IWRM ascended to near hegemony in the water sector in the 1990s (see Conca 2006). A related concern was that attention to the environment—and of thinking about the environment in global terms—could be hedged in by IWRM, which focused on the watershed scale and not on global concerns. Indeed, by the end of the millennium key figures in global hydrology, such as Malin

Falkenmark (2001), increasingly pressed the global community to link water, food, and the environment in broader discussions regarding the governance of scarce resources.

The focus on watershed concerns in IWRM was part of a long-standing process of identifying the river basin, or watershed, as the natural unit of management (Cohen and Davidson 2011). The idea was intuitive: Since water is a key environmental, industrial, and social constraint and since watersheds are the spatial units capturing and directing water flows, the watershed presents a seemingly natural scale for decision making. The assumption, though widely touted, has serious flaws in terms of governance (see Warner et al. 2008): One is that determining the scale of the "watershed" is a political decision given that many large watersheds, such as the Nile or the Ganges, have many sub-basins that are both large and complicated. A second is that many institutional and jurisdictional boundaries do not follow those of watersheds, so judgments will inevitably be made regarding how this "natural" unit is defined. Particularly in international contexts, where water routinely crosses national borders, the claimed "naturalness" of watersheds can only be maintained by highly orchestrated governance arrangements (Blatter and Ingram 2001). Nevertheless, the watershed remains an important spatial unit for governance as it has been for management because, even despite its fuzzy political edges, it is often the scale at which science, policy, and politics intersect on environmental issues (Cohen 2012).

The Second World Water Forum in The Hague in 2000 proved a turning point for water governance, which was identified as the first of three priority action areas at the International Conference on Freshwater in Bonn in 2001 (Rogers and Hall 2003). That year also marked the start of the Millennium Ecosystem Assessment, a global collaboration of scientists that published its results in 2005 and which proved an important influence in the shift toward resilience-based approaches to environmental management and governance. As Chap. 1 covered, by the early 2000s there were several tensions within the global water community over how to achieve IWRM, yet there was also considerable momentum as a growing number of countries adopted IWRM as part of various development projects. Amid this tension, resilience-based approaches to environmental management and governance offered a unique solution that left the policy frameworks of IWRM in place while providing a new perspective on the complex and changing ecological conditions faced by water managers. Resilience refers to the capacity of a socio-ecological system to adapt to disturbances that result from either human or non-human forces while still maintaining its functions and feedbacks (see Folke 2006). In the new millennium, it presented a way to maintain policy stability in places where IWRM was adopted while also opening the path for understandings of "integration" to migrate from one premised on the "balance of nature" to one in which constant change and coevolution were the norm.

The cultivation of resilience-based, adaptive management techniques alongside IWRM policies presented a significant moment for aligning water management with the approach to global environmental governance that appreciated the nonlinear and multiscalar dynamics of complex systems. It also provided a certain amount of flexibility as the experimental approaches of adaptive management could be used to legitimate the testing and trial of new governance structures for sustainable development (Feldman 2007). Conceptualizing and reinvigorating IWRM through adaptive management techniques involved expanding the traditional focus on water management from the "blue" water of lakes and rivers to incorporate the links of water management to "green" water flows, such as evaporation, that were closely tied to land-cover change and ecosystem processes (Falkenmark and Rockström 2004). When the Millennium Ecosystem Assessment (2005) released its findings in 2005, the shift toward learning about water governance through techniques of adaptive management strengthened the connections between shared approaches to understanding both water governance and global environmental change (Pahl-Wostl 2007). The findings of the Millennium Ecosystem Assessment further bolstered the momentum that the first WWDR had achieved in 2003 by once more emphasizing the deep links between humans and the Earth system. The 2003 WWDR report, *Water for People, Water for Life*, argued that the global water crisis had deep social and environmental interconnections and precipitated the UN Declaration of the *Water for Life* decade in 2005.

A critical component to the resilience of socio-ecological systems was the need to preserve adequate environmental flows for ecosystem functioning. Calculating the water flows needed for different aquatic ecosystems to remain healthy over time, space, and scale has been attempted through a variety of methodological approaches (see Acreman 2016). All of these are part of the growing appreciation of human impacts on aquatic ecosystems that led scientists and policy makers, including environmental NGOs and the World Bank, to begin to take concerns over environmental flows more seriously (Dyson et al. 2003; King and Brown 2006; Hirgi and Davis 2009). In Europe, the EU Water Framework Directive became a key piece of policy for regulating environmental flows within and among member states (see Acreman et al. 2008). The growing recognition of the need for enhanced environmental flows, however, faced steep challenges not only from years of entrenched water use practices—many codified in legal rights—but also from the more severe and intense extremes of water resulting from climate change. In Europe, heavily developed rivers faced the prospect of increased intensity of rainfall and flooding (Christensen and Christensen 2003). In Africa, climate-induced flood events exposed deep social inequalities in the development of infrastructure, especially among the urban poor (Douglas et al. 2008). What was becoming clear was that environmental flows did not equate to natural flows and that, going forward, the human imprint on the global water system would need to be dealt with at multiple scales and in the context of existing, often unequal social structures.

The 2006 WWDR, *Water a Shared Responsibility*, further advanced the theme of social inequality and the need for integrated approaches to water to move away from business as usual. The report engaged with a wide spectrum of pathways to solve water challenges, including addressing water and sanitation services, agriculture, energy, and a specific focus on environmental sustainability (UNESCO 2006). The report captured in governance terms the increasing complexity besetting global water challenges as human demands on hydrological systems continued to grow. Throughout the first decade of the new millennium, a spate of studies revealed human impacts on freshwater ecosystems while new technologies, such as the paired GRACE satellites, provided new technologies for assessing groundwater (Meybeck 2003; Vörösmarty et al. 2004, 2010; Famiglietti et al. 2011). Increasingly, scholars and practitioners sought ways to connect local and global water challenges in the Anthropocene—the new epoch coined to capture the many ways in which humans now rival the great forces of nature (see Gupta et al. 2013). Recognition of these challenges was met by shifts in the hydrological sciences to more fully assess human impacts on the global water system at multiple scales and in a context where water was one component of a set of interlocking Earth system dynamics (Vörösmarty et al. 2013; Savenije et al. 2014).

With the growing recognition of complexity and change in human-water relationships, it became increasingly clear that no single framework for governance could manage water in a holistic sense. Or did it? The rejection of IWRM and its particular approach to holism also spurred a new approach to holism—where, rather than try to assemble all the dynamics affecting water into a single framework, the aim is to govern the overall dynamic of an interconnected system in which water is connected to food, energy, and the climate. One way to contrast the two approaches is to note that IWRM sought to *collect* all of the variables affecting water as the basis for integrated, holistic decision making. By contrast, what is now known as the water–energy–food–climate nexus seeks to govern the *connections* among water and multiple other sectors. This type of holism is still concerned with the global water system but it is not fixated on making water the common denominator for integration. Governing these multiple connections became a central element of environmental concern in 2011 at a conference in Bonn in preparation for the Rio+20 conference set for 2012 (Hoff 2011). There are many elements to the emerging discourse on governance and the nexus (see Chap. 3), but one of the key features of this new approach is the attempt to reconcile the fact that environmental impacts on water can come from multiple—often unanticipated—directions in an era of global environmental change (Pittock et al. 2015).

The turn to the nexus is part of a broad shift currently underway in global water governance toward thinking about water as part of a set of interconnected social-ecological systems. It is a shift with important implications for IWRM (see Benson et al. 2015) and also one being affected by new governance institutions, such as the Global Water System Project (2014). Here, once more, it is critical to highlight the ways in which numerous values come to bear on how science is linked to policy, especially when recognition of interlinked Earth and economic systems is brought into discussions and policies of sustainability (Hussey and Pittock 2012; Lawford et al. 2013; Ringler et al. 2013). There is a concern, both material and moral, that the exchange of one set of scientific ideas about the global hydrological cycle for those of Earth system sciences may still connect governance concerns to the environment in ways that exclude or marginalize non-western perspectives (Schmidt et al. 2016; Schmidt 2017a). To see how the evolution of approaches to water management and water governance can operate in this way, it is helpful to consider a case study of the Mekong River Basin, where numerous interventions over several decades reveal the difficulties related to environmental aspects of global water governance. There, and despite the laudable principles of good governance regarding participation and transparency, little or no space for participation is actually provided, while the environment is considered secondary, or not all, in comparison with economic growth and development. Across the Mekong Basin, as in much of the developing world, the space for an environmental voice in water governance is often significantly curtailed by geopolitical and economic development agendas.

THE MEKONG RIVER

The 800,000 km² Mekong Basin is home to the Mekong River, the seventh longest river in the world. The transboundary Mekong River is shared by six countries: China, Burma, Thailand, Laos, Cambodia, and Vietnam. Known as "the breadbasket of Southeast Asia," the Mekong Basin is home to a population of 70 million with 90 distinct ethnic groups (Galipeau et al. 2013). With its extensive wetlands and floodplains, the Basin supports the largest inland fisheries in the world with an annual catch of 2.6 million tonnes and over 500,000 tonnes of other aquatic animals (e.g., aquatic insects, amphibians, and mollusks) valued annually at between $3.9 and $7 billion (Hortle 2007). Over two-thirds of the Basin's population are involved in fishing for their livelihoods or to support food security (Mekong River Commission 2003). In the Lower Mekong Basin, aquatic resources make up between 47 and 80% of animal protein in rural diets (Baran and Ratner 2007; Friend and Blake 2009). Biodiversity in the Mekong Basin is second only to the Amazon, with over 1200 species of fish and a number of endemic and endangered species such as the giant Mekong catfish, which can grow to three meters and can weigh over 300 kg, the giant Mekong stingray, and the Mekong river dolphin (Dugan et al. 2010). Although a healthy and functioning environment in the Mekong has been shown to underpin the livelihoods and food security of much of the basin's population, the agenda and unfolding of water governance and the space for environmental considerations have been largely a western import that has faced obstruction by regional and local agendas seeking rapid economic growth through hydropower development, transport, and irrigation.

An early contribution to the emergence of environmental water governance in the Mekong was a Ford Foundation-sponsored report by

Gilbert White (White et al. 1962) entitled, *Economics and Social Aspects of Lower Mekong Development.* The White Report went beyond the engineering and technical considerations of previous studies into water resource management across the basin to look at the potential environmental and social impacts of development. For White et al. (1962), the report drew from concerns emerging in the growing environmental movement in the USA that was shifting the development purview beyond a quest for growth at any cost to encompass considerations of environmental impacts and the integrated nature of water, humans, and the environment. The report was also a way for White to illustrate Asia's first, large-scale efforts to study the economic, institutional, and social aspects of development prior to development actually occurring.

After a period of political unrest and conflict across the Basin, the coordinated, western-led effort behind water governance linked with the environment was picked up again in 1995 with the formation of the Mekong River Commission (MRC). The MRC is primarily a western-funded River Basin Organization (RBO) representing all the states in the Basin, but with China and Myanmar only acting as dialogue partners. The MRC funding donors introduced a set of governance objectives that moved away from the heavy focus on water resources development to one that contained principles based on "sustainable development, utilisation, management and conservation of the water and related resources of the Mekong River Basin" (MRC 1995).

The source of the MRC's funding in western institutions has shaped the emergence and dissemination of good water governance principles such as participation and transparency, and the consideration of environmental impacts of development. These environmental water governance principles imported by the MRC have been further espoused by western-funded INGOs. Under the MRC, this new development agenda emphasized cooperation around scientific studies, capacity building, and environmental protection (Jacobs 2002). For its donors, the MRC also provides a strategic opportunity to open dialogue spaces with the region's emerging markets. Finally, it allows them to meet aid objectives by encouraging good governance principles in the Basin's future water management (McCawley 2001).

For some governments in the region, these new principles were not well suited to their existing development pathways or political systems. The government of Laos, for example, may have perceived the concept of good water governance as a threat to its power base through the promotion of devolution from the government to a form of governance that included the state, the market, civil movements (NGOs), and civil society (Matthews and Schmidt 2014). As noted by Ribot (2004), the devolution of power is often met by strong resistance from those in power. Although the Lao government was reluctant to accept these new principles, the MRC and its donors were also seen as an important source of much needed funding and economic stimulus for underdeveloped states. With increased confidence and stability in the region, from 1990 to 1995, net Overseas Development Assistance flows to Thailand and Indochina rose by approximately 400% from $422 million to $1.66 billion USD (OECD 1997).

The MRC was not the only funder espousing the principles of environmental water governance. From the 1990s, The Asian Development Bank (ADB) and the World Bank also promoted a brand of good water governance, but one that was more closely aligned with neoliberal policies that encouraged the market-led development of natural resources recognizing that environmental impacts were part of the trade-offs needed to develop. A centerpiece of this agenda was the implementation of the Greater Mekong Subregion (GMS) Programme, a scheme strongly focused on the connectivity of markets and economies and private sector investment in hydropower development to advance economic growth and reduce poverty within the framework of good governance (Middleton et al. 2009). The GMS mandate has promoted interconnectivity and hydropower development, including private sector investment in mainstream and tributary dams, but still included calls for participation and transparency in decision making although with less emphasis on preserving ecosystem health (Cornford and Matthews 2007). The program's focus on development appears to have been more rapidly accepted by the region with all basin states signing on as members.

Despite good governance principles being strongly promoted by the MRC, the ADB, and the World Bank, a scalar disconnect has arisen between these agendas and those of many of the basin states. This disconnect may have emerged when the member states assumed the MRC would follow a development path that pursued rapid economic growth similar to the path taken in the USA and Europe through the construction of mainstream and tributary hydropower dams and large irrigation projects (Lang 2005). The disconnect between the environmental and water governance agendas of western donors and NGOs in the region and that of basin states has emerged as a key point of contention within the MRC. Ultimately, the deadlock around this debate has nullified the meaningful implementation of many environmental water governance principles across the basin. As a result, the concerns of international donors for holistic, participatory water management, programs of environmental and ecosystem protection, and monitoring and evaluation have been tolerated by member states to demonstrate their commitment to these processes, but at the same time government policies have continued to focus on top–down, non-transparent decision making and rapid water resource development despite its impacts on ecosystems (Suhardiman et al. 2012). In fact, to date the MRC has largely only managed to gain cooperation from all of its member states on apolitical issues. This disconnect allows governments to implement policies of sovereign interest "because the MRC lacks power to direct transboundary water governance issues in the region" (ibid).

The disconnect between competing agendas and differing interpretations of governance and environment valuation and protection has delegitimized other efforts at IWRM and basin-wide water management plans, further increasing the lack of transparency. National governments may view any efforts introduced by outside actors concerned with the environment or social issues as something they must pay lip service to, but which are counter to national efforts to quickly modernize or develop water resources (Geheb et al. 2014).

Meanwhile, the World Bank and ADB's assistance with domestic policies and the GMS Programme have scaled good governance principles alongside a neoliberal vision of rapid economic growth including hydropower and irrigation development. This has shifted the concept of participation and transparency away from individual countries and local scales to a larger scale that creates an area of perceived harmonious community called the "Greater Mekong Subregion." This scaling of participation and transparency at a basin level has meant that the application of good water governance principles including environmental considerations can be achieved through INGO participation in planning although this participation may be tokenistic (Matthews and Schmidt 2014). In reality, many basin states have only engaged with non-local actors and avoided any real devolution of power, thereby creating an illusion that they are following the norms of IWRM and good water governance promoted by donors.

INGOs across the Mekong Basin have also used water governance as a vehicle to articulate power and legitimacy through their western representation and the discourses and practice they promote (Peet and Watts 2004; Matthews 2012). This perceived legitimacy and western-influenced knowledge may be both unrepresentative of local needs and clashing with regional development discourses. For example, many INGOs in the region promote a brand of IWRM and water governance that heavily critiques hydropower as being devastating to the environment of the region while downplaying its economic benefits. Sundberg (1998, p. 14) argues that they assume "the moral authority to speak for nature." This use of western knowledge by INGOs enables them to articulate a form of power and claim legitimacy within environmental and social impact debates, but is out of step with the development agenda of regional states. This INGO perspective has also not always been accurately representative of local level needs (Sunderberg 1998). This clash of agendas and forms of knowledge has created tensions between states, the private sector, and INGOS. As a result, in many cases states have delimited their influence and controlled their activities.

The above examples illustrate how different agendas shape the application of water governance to environmental concerns. Increasing pressure on water due to climate change and population growth can also have detrimental impacts on the establishment of water governance. In 2016, for example, an El Niño-inspired drought across the Mekong Basin resulted in unilateral decisions around water management with Thailand installing dozens of pump stations along the Mekong mainstream, seeking to withdraw 47 million cubic meters of water over 3 months to irrigate thirsty fields in the country's northeast—one of the country's poorest regions. Downstream countries asked that Thailand's activities be submitted to scrutiny by the MRC as part of a process of water governance. Thailand, however, rejected the request due to the pressure to appease domestic politics.

Across the basin, the MRC's dominance of the water governance agenda has been under pressure. The MRC's legitimacy has been under question due to the many deep divisions between the Mekong countries' growth agendas and that of western donors' water governance. The institution has been further wracked by reforms and significant changes to its modus operandi, which only serves to deepen the ambiguities associated with regional water governance and management. The MRC has found that with increasing demands for energy and water for irrigation and the deep inter-linkages of water to political economies across the region, it is progressively difficult for a RBO to build cooperation and a shared vision for the development of the Mekong Basin. Hale and Held (2012, p. 169) articulate this challenge in their analysis of global governance when they state that "...the emergence of new powers, the growing challenges of collective action problems and the complexity of institutions that seek to address them—have made it increasingly difficult to govern transnational problems through multilateral cooperation." Although the MRC may find these new dynamics increasingly constraining, the new governance reality offers space for non-traditional actors such as the private sector, to influence, either positively or negatively, the ability of the states to promote good water governance practices.

As a result of this weakening of the MRC, in March 2016, China announced the creation of the Lancang-Mekong Cooperation Mechanism (LMCM). The LMCM is designed to "strengthen cooperation in such fields as infrastructure, engineering machinery, electricity, construction materials and communications." The mechanism includes a RMB 10 billion yuan concessional loan and a US$10 billion credit line, including a US$5 billion preferential export buyers' credit and a US$5 billion special loan on production capacity cooperation. Its members include all the regional riparian states (unlike the MRC, in which China and Myanmar are only dialogue partners). The fund aims to finance up to 20 dams in the region. Chinese firms are already the most active dam builders in the Mekong with 57 dams commissioned, 12 under construction, and 16 planned or proposed in the region (Matthews and Motta 2015). The latter includes the soon-to-be-announced Pak Beng on the Mekong mainstream in Laos, to be developed by Datang International.

The LMCM is more than just finance, however. The mechanism's focus on cooperation over water resources increases China's hegemony in the region and could potentially undermine the efforts of the western-funded MRC. The mechanism also increases China's bargaining power and influence on the lower basin states. Through the LCMC, China can exert influence on ASEAN members on issues such as the South China Sea. The LCMC allows China to lead the direction and set the tone of water governance across the region.

Both China and the USA have been working to influence the emergence of water governance in Myanmar and along the Irrawaddy and Salween rivers: The 6000 megawatt Myitsone Dam, which was to be developed by the China Power Investment Corporation, was suspended in 2011. Its suspension by the Myanmar president had a significant impact on China's relationship with its southern neighbor. In September 2016, fighting in Karen was said (by activist and NGO groups) to be an effort by the Burmese Border Guard Force to clear people from the Hatgyi Dam's inundation zone. This dam is slated for development by Sinohydro, and the forced displacement has fueled speculation that the Myanmar military is working closely with Chinese hydropower development companies. The World Bank has given a US$100 million loan to strengthen and improve Myanmar's water management capabilities, thereby ensuring that US-led interests in water governance across the region are also prominent.

As in many parts of the world, the basin states of the Mekong have been consumers of international development knowledge. The principles of participation, equity, environmental sustainability, and transparency in water governance and decision making have been imported and driven into the region from outside agendas. The introduction of the good governance principles drew from an international trend encouraging a new "softer development agenda" focusing on good governance issues that emerged in the 1980s and one that was heavily promoted by INGOs from the mid-1990s (McCawley 2001). As the Mekong Basin case demonstrates, however, how the western-led and imported agendas of water governance and environmental concerns are interpreted and applied by countries remains contested. The Mekong case study is representative of the close linkages between water governance, economics, and the power of environmental concerns. As we move deeper into the increasingly unpredictable Anthropocene, the pressures of population growth, consumption, climate change, and the associated increased water demand will make water governance and the management of environmental trade-offs an increasingly contested space.

CONCLUSION

The Mekong River case provides a snapshot of the intersections of western environmental values, global governance, and the dynamics at play in linking science to policy. It is also emblematic of the politics and economics that shape the discourse around water governance in much of the developing world. Increasingly, calls to shift toward a "green economy" for the purposes of sustainable development are met skeptically by developing nations who fear that, under the pretense of environmental policy, they will be subject to new and unfavorable economic arrangements

(Conca 2015). It is a legitimate concern given that international development has, despite constant claims to the contrary, facilitated a massive shift in wealth from the Global South to the Global North (Kar and Schjelderup 2016). These practices have not infrequently exacerbated environmental harms in ways that reflect and reinforce the prevailing inequalities of social and political structures (Sachs 1999; Downey 2015).

Nevertheless, global water governance is critically dependent on a shared understanding of the environment as a way to link the many facets and demands of humans, non-humans, and ecological processes on water. Over the decades since Mar del Plata, there has been an increasingly sophisticated accounting of human impacts on the global water system and the regional and local causes of pressures at lower scales. How these scientific assessments fit with the institutions and programs of global water governance, however, has been heavily influenced by western values and concerns. Notably, US- and UN-led networks and organizations have supported agendas of IWRM that captured international water management that later came to dominate global water governance. The upshot of having western ideas and norms facilitate the links between scientific assessments of the global water system and programs of global water governance is that these values have similarly pervaded approaches to other facets of sustainability regarding economies and societies.

References

Acreman, Mike. 2016. Environmental Flows—Basics for Novices. *WIREs Water* 3 (5): 622–628.

Acreman, M., M. Dunbar, J. Hannaford, O. Mountford, P. Wood, N. Holmes, I. Cowx, R. Noble, C. Extence, J. Aldrick, J. King, A. Black, and D. Crookall. 2008. Developing Environmental Standards for Abstractions from UK Rivers to Implement the EE Water Framework Directive. *Hydrological Sciences* 53 (6): 1105–1120.

Ahmed, Izrar, Abdulaziz A. Al-Othman, and Rashid Umar. 2014. Is shrinking Groundwater Resources Leading to Socioeconomic and Environmental Degradation in Central Ganga Plain, India? *Arabian Journal of Geosciences* 7 (10): 4377–4385.

Alley, William M., and Rosemarie Alley. 2017. *High and Dry: Meeting the Challenges of the World's Dependence on Groundwater.* New Haven: Yale University Press.

Baran, Eric, and Blake Ratner. 2007. *The Don Sahong Dam and Mekong Fisheries.* Science Brief 3. Penang: WorldFish Center.

Benson, David, Animesh K. Gain, and Josselin J. Rouillard. 2015. Water Governance in a Comparative Perspective: From IWRM to a 'Nexus' Approach? *Water Alternatives* 8 (1): 756–773.

Birkenholtz, Trevor L. 2015. Recentralizing Groundwater Governmentality: Rendering Groundwater and Its Users Visible and Governable. *WIREs Water* 2: 21–30.

Biswas, Asit K. (ed.). 1978. *United Nations Water Conference: Summary and Main Documents.* Oxford: Pergamon Press.

Biswas, Asit K. 2004. From Mar Del Plata to Kyoto: A Review of Global Water Policy Dialogues. *Global Environmental Change* 14: 81–88.

Biswas, Asit K., and Ceilia Tortajada (eds.). 2009. *Impacts of Megaconferences on the Water Sector.* New York: Springer.

Blatter, Joachim, and Helen Ingram (eds.). 2001. *Reflections on Water: New Approaches to Transboundary Conflict and Cooperation.* Cambridge, MA: Institute of Technology.

Carpenter, Stephen R., Emily H. Stanley, and M. Jake Vander Zanden. 2011. State of the World's Freshwater Ecosystems: Physical, Chemical, and Biological Changes. *Annual Review of Environment and Resources* 36: 75–99.

Christensen, Jens H., and Ole B. Christensen. 2003. Climate Modelling: Severe Summertime Flooding in Europe. *Nature* 421: 805–806.

Cohen, Alice. 2012. Rescaling Environmental Governance: Watersheds as Boundary Objects at the Intersections of Science, Neoliberalism, and Participation. *Environment and Planning A* 44 (9): 2207–2224.

Cohen, Alice, and Seanna Davidson. 2011. The Watershed Approach: Challenges, Antecedents, and the Transition from Technical Tool to Governance Unit. *Water Alternatives* 4 (1): 1–14.

Conca, Ken. 2006. *Governing Water: Contentious Transnational Politics and Global Institution Building.* Cambridge: MIT Press.

Conca, Ken. 2015. *An Unfinished Foundation: The United Nations and Global Environmental Governance.* Oxford: Oxford University Press.

Cornford, Jonathan, and Nathanial Matthews. 2007. *Hidden Costs: The Underside of Economic Transformation in the Greater Mekong Subregion.* Carlton: Qxfam Australia.

Dalin, Carole, Yoshihide Wada, Thomas Kastner, and Michael J. Puma. 2017. Groundwater Depletion Embedded in International Food Trade. *Nature* 543: 700–704.

Deemer, Bridget R., John A. Harrison, Siyue Li, J. Jake Beaulieu, Tonya DelSontro, Nathan Barros, José F. Bessera-Neto, Stephen M. Powers, Marco A. dos Santos, and J. Arie Vonk. 2016. Greenhouse Gas Emissions from Reservoir Water Surfaces: A New Global Synthesis. *BioScience* 66 (11): 949–964.

Douglas, Ian, M. Kurshid Alam, Y. Maghenda, L. Mclean Mcdonnell, and J. Campbell. 2008. Unjust Waters: Climate Change, Flooding and the Urban Poor in Africa. *Environment & Urbanization* 20 (1): 187–205.

Downey, Liam. 2015. *Inequality, Democracy, and the Environment*. New York: New York University Press.

Dugan, Patrick J., Chris Barlow, Angelo A. Agostinho, Eric Baran, Glenn F. Cada, Daqing Chen, Ian G. Cowx, et al. 2010. Fish Migration, Dams, and Loss of Ecosystem Services in the Mekong Basin. *AMBIO: A Journal of the Human Environment* 39 (4): 344–348.

Dyson, M., G. Bergkamp, and J. Scanlon (eds.). 2003. *Flow: The Essentials of Environmental Flows*. Cambridge: International Union for Conservation of Nature.

Ekbladh, David. 2010. *The Great American Mission: Modernization and the Construction of an American World Order*. Princeton: Princeton University Press.

Escobar, Arturo. 2008. *Territories of Difference: Movements, Life, Redes*. Durham: Duke University Press.

Escobar, Arturo. 2012. *Encountering Development: The Making and Unmaking of the Third World*. Princeton: Princeton University Press.

Espeland, Wendy N. 1998. *The Struggle for Water: Politics, Rationality, and Identity in the American Southwest*. Chicago: University of Chicago Press.

Falkenmark, Malin. 2001. The Greatest Water Problem: The Inability to Link Environmental Security, Water Security and Food Security. *International Journal of Water Resources Development* 17 (4): 539–554.

Falkenmark, Malin, and Carl Folke. 2010. Ecohydrosolidarity: A New Ethics for Stewardship of Value-Adding Rainfall. In *Water Ethics: Foundational Readings for Students and Professionals*, ed. Peter G. Brown and Jeremy J. Schmidt, 247–264. Washington, DC: Island Press.

Falkenmark, Malin, and Johan Rockström. 2004. *Balancing Water for Humans and Nature: The New Approach in Ecohydrology*. London: Earthscan.

Famiglietti, J.S., M. Lo, S. Ho, J. Anderson, J. Bethune, T. Syed, S. Swenson, C. de Linage, and M. Rodell. 2011. Satellites Measure Groundwater Depletion in California's Central Valley. *Geophysical Research Letters* 38: L03403.

Fearnside, Philip M. 2016. Environmental and Social Impacts of Hydroelectric Dams in Brazilian Amazonia: Implications for the Aluminum Industry. *World Development* 77: 48–65.

Feldman, David. 1995. *Water Resources Management: In Search of an Environmental Ethic*. Baltimore: John Hopkins University Press.

Feldman, David. 2007. *Water Policy for Sustainable Development*. Baltimore: John Hopkins University Press.

Feldman, David L. 2012. *Water*. Cambridge: Polity Press.

Folke, Carl. 2003. Freshwater for Resilience: A Shift in Thinking. *Philosophical Transactions of the Royal Society of London B* 358: 2027–2036.

Folke, Carl. 2006. Resilience: The Emergence of a Perspective for Social-Ecological Systems Analyses. *Global Environmental Change* 16: 253–267.

Friend, Richard M., and David J.H. Blake. 2009. Negotiating Trade-offs in Water Resources Development in the Mekong Basin: Implications for Fisheries and Fishery-Based Livelihoods. *Water Policy* 11 (S1): 13–30.

Gaard, Greta. 2001. Women, Water, Energy: An Ecofeminist Approach. *Organization & Environment* 14 (2): 157–172.

Galipeau, Brendan A., Mark Ingman, and Bryan Tilt. 2013. Dam-Induced Displacement and Agricultural Livelihoods in China's Mekong Basin. *Human Ecology* 41 (3): 437–446.

Gardner, Royal C., Stefano Barchiesi, Coralie Beltrame, C.M. Finlayson, Thomas Galewski, Ian Harrison, Marc Paganini, et al. 2015. *State of the World's Wetlands and Their Services to People: A Compilation of Recent Analyses.* Gland: Ramsar Convention Secretatiat.

Geheb, Kim, Niki West, and Nathanial Matthews. 2014. The Invisible Dam: Hydropower and Its Narration in the Lao People's Democratic Republic. In *Hydropower Development in the Mekong Region: Political, Socio-Economic and Environmental Perspectives*, ed. Nathanial Matthews and Kim Geheb, 101–126. London: Routledge.

Gleick, Peter H. 1989. Climate Change, Hydrology, and Water Resources. *Review of Geophysics* 27 (3): 329–344.

Gleick, Peter H. (ed.). 1993. *Water in Crisis: A Guide to the World's Fresh Water Resources.* New York: Oxford University Press.

Gleick, Peter H., and Meena Palaniappan. 2010. Peak Water Limits to Freshwater Withdrawal and Use. *Proceedings of the National Academy of Sciences* 107 (25): 11155–11162.

Global Water System Project. 2014. Call to Action for Implementing the Water-Energy-Food Nexus. International Conference: Sustainability in the Water-Energy-Food Nexus, 19–20 May 2014 in Bonn, Germany, The Global Water System Project.

Guha, Ramachandra. 2000. *Environmentalism: A Global History.* New York: Longman.

Gupta, Joyeeta, Claudia Pahl-Wostl, and Ruben Zondervan. 2013. 'Glocal' Water Governance: A Multi-Level Challenge in the Anthropocene. *Current Opinion in Environmental Sustainability* 5 (6): 573–580.

Hale, Thomas, and David Held. 2012. Gridlock and Innovation in Global Governance: The Partial Transnational Solution. *Global Policy* 3 (2): 169–181.

Harding, Sandra. 2015. *Objectivity and Diversity: Another Logic of Scientific Research.* Chicago: University of Chicago Press.

Hirgi, R., and R. Davis. 2009. *Environmental Flows in Water Resources Policies, Plans, and Projects: Findings and Recommendations.* Washinton, DC: The World Bank.

Hoff, Holger. 2011. *Understanding the Nexus. Background Paper for the Bonn 2011 Conference: The Water Energy and Food Security Nexus*. Stockholm: Stockholm Environment Institute.

Holling, C.S. 1986. The Resilience of Terrestrial Ecosystems: Local Surprise and Global Change. In *Sustainable Development of the Biosphere*, ed. William C. Clark and R.E. Munn, 292–316. Cambridge: Cambridge University Press.

Hortle, Kent G. 2007. Consumption and the Yield of Fish and Other Aquatic Animals from the Lower Mekong Basin. *MRC Technical Paper* 16: 1–88.

Hussey, Karen, and Jamie Pittock. 2012. The Energy-Water Nexus: Managing the Links Between Energy and Water for a Sustainable Future. *Ecology and Society* 17 (1): 31.

Huxley, Julian. 1935. Plans for Tomorrow: The Tennessee Valley Authority. *The Listener*, November 20, 897–900.

Huxley, Julian. 1943. *TVA: Adventure in Planning*. Cheam, Surrey: The Architectural Press.

Huxley, Julian. 1946. *UNESCO: Its Purpose and Philosophy*. Paris: UNESCO.

Jacobs, Jeffrey W. 2002. The Mekong River Commission: Transboundary Water Resources Planning and Regional Security. *The Geographical Journal* 168 (4): 354–364.

Jones-Lepp, Tammy L., Charles Sanchez, David A. Alvarez, Doyle C. Wilson, and Randi-Laurant Taniguchi-Fu. 2012. Point Sources of Emerging Contaminants Along the Colorado River Basin: Source Water for the Arid Southwestern United States. *Science of the Total Environment* 430: 237–245.

Kar, Dev, and Guttorm Schjelderup. 2016. *Financial Flows and Tax Havens: Combining to Limit the Lives of Billions of People*. Washington, DC: Global Financial Integrity.

Kennedy, T.A., W.F. Cross, R.O. Hall Jr., C.V. Baxter, and E.J. Rosi-Marshall. 2013. *Native and Nonnative Fish Populations of the Colorado River Are Food Limited—Evidence From New Food Web Analyses: US Geological Survey Fact Sheet 2013–3039*, 4 pp. https://pubs.usgs.gov/fs/2013/3039/.

King, J., and C. Brown. 2006. Environmental Flows: Striking the Balance Between Development and Resource Protection. *Ecology and Society* 11 (2): Art. 26.

Korzoun, V.I., A.A. Sokolov, M.I. Budyko, K.P. Voskresensky, G.P. Kalinin, A.A. Konoplyantsev, E.S. Korotkevich, and M.I. Lvovich (eds.). 1978. *Atlas of World Water Balance: Water Resources of the Earth*. Paris: UNESCO.

L'vovich, Mark I., and Gilbert F. White. 1990. Use and Transformation of Terrestrial Water Systems. In *The Earth as Transformed by Human Action: Global and Regional Changes in the Biosphere Over the Past 300 Years*, ed. B.L. Turner II, 235–252. Cambridge: CUP Archive.

Lang, Malee T. 2005. Management of the Mekong River Basin: Contesting its Sustainability from a Communication Perspective. Working Paper 130, Development Research Series.

Lautze, Jonathan, Sanjiv De Silva, Mark Giordano, and Luke Sanford. 2011. Putting the Cart Before The Horse: Water Governance and IWRM." *Natural Resources Forum* 35 (1): 1–8 (Blackwell Publishing Ltd.).

Lawford, Richard, Janos Bogardi, Sina Marx, Sharad Jain, Claudia Pahl-Wostl, Kathrin Knüppe, Claudia Ringler, Felino Lansigan, and Francisco Meza. 2013. Basin Perspectives on the Water-Energy-Food Security Nexus. *Current Opinion in Environmental Sustainability* 5 (6): 607–616.

Lazarus, Richard J. 2004. *The Making of Environmental Law.* Chicago: The University of Chicago Press.

Lehner, Bernhard, C. Reidy Liermann, Carmen Revenga, Charles J. Vörösmarty, Balazs Fekete, Philippe Crouzet, Petra Döll, Marcel Endejan, Karen Frenken, Jun Magome, Christer Nilsson, James C. Robertson, Raimund Rödel, Nikolai Sindorf, and Dominik Wisser. 2011. High-Resolution Mapping of the World's Reservoirs and Dams for Sustainable River-Flow Management. *Frontiers in Ecology and the Environment* 9 (9): 494–502.

Lélé, Sharachchandra M. 1991. Sustainable Development: A Critical Review. *World Development* 19 (6): 607–621.

Linton, Jamie. 2010. *What is Water? The History of a Modern Abstraction.* Vancouver, BC: UBC Press.

Maavara, Taylor, Ronny Lauerwald, Pierre Regnier, and Phillippe van Cappellen. 2017. Global Perturbation of Organic Carbon Cycling By River Damming. *Nature Communications* 8: 15347.

Macekura, Stephen J. 2015. *Of Limits and Growth: The Rise of Global Sustainable Development in the Twentieth Century.* Cambridge: Cambridge University Press.

Matthews, Nathanial. 2012. Water Grabbing in the Mekong Basin—An Analysis of the Winners and Losers of Thailand's Hydropower Development in Lao PDR. *Water Alternatives* 5 (2): 392–411.

Matthews, Nathanial. 2016. People and Fresh Water Ecosystems: Pressures, Responses and Resilience. *Aquatic Procedia* 6: 99–105.

Matthews, Nathanial, and Jeremy J. Schmidt. 2014. False Promises: The Contours, Contexts and Contestation of Good Water Governance in Lao PDR and Alberta, Canada. *International Journal of Water Governance* 2 (2/3): 21–40.

Matthews, Nathanial, and Stew Motta. 2015. Chinese State-Owned Enterprise Investment in Mekong Hydropower: Political and Economic Drivers and Their Implications Across the Water, Energy, Food Nexus. *Water* 7 (11): 6269–6284.

McCawley, Peter. 2001. Asian Poverty: What Can Be Done? Discussion Paper No. 292, School of Economics, University of Queensland.

McDonald, Robert I., Katherine Weber, Julie Padowski, Martina Flörke, Christof Schneider, Pamela A. Green, Thomas Gleeson, Stephanie Eckman, Bernhard Lehner, Deborah Balk, Timothy Boucher, Günther Grill, and Mark

Montgomery. 2014. Water on an Urban Planet: Urbanization and the Reach of Urban Water Infrastructure. *Global Environmental Change* 27: 96–105.

McNeill, J.R., and Peter Engelke. 2016. *The Great Acceleration: An Environmental History of the Anthropocene Since 1945.* Cambridge: Harvard University Press.

Mekong River Commission. 1995. *Agreement on the Cooperation for the Sustainable Development of the Mekong River Basin,* 5 April 1995. Mekong River Commission.

Mekong River Commission. 2003. *State of the Basin Report.* Vientiane: Mekong River Commission.

Merchant, Carolyn. 1980. *The Death of Nature: Women, Ecology, and the Scientific Revolution.* San Francisco: Harper & Row.

Meybeck, Michel. 2003. Global Analysis of River Systems: From Earth System Controls to Anthropocene Syndromes. *Philosophical Transactions of the Royal Society B* 358: 1935–1955.

Middleton, Carl, Jelson Garcia, and Tira Foran. 2009. Old and New Hydropower Players in the Mekong Region: Agendas and Strategies. *Contested Waterscapes in the Mekong Region: Hydropower, Livelihoods and Governance,* ed. François Molle and Tira Foran, 23–54. London: Earthscan.

Millennium Ecosystem Assessment. 2005. *Ecosystems and Human Well-Being: Wetlands and Water Synthesis.* Washington, DC: World Resources Institute.

Milly, P.C.D., J. Betancourt, M. Falkenmark, R.M. Hirsch, Z.W. Kundzewicz, D.P. Lettenmaier, and R.J. Stouffer. 2008. Stationarity is Dead: Whither Water Management? *Science* 319: 573–574.

Mitchell, Timothy. 2014. Economentality: How the Future Entered Government. *Critical Inquiry* 40 (4): 479–507.

Nace, Raymond L. 1969. *Water and Man: A World View.* Paris: UNESCO.

Nace, Raymond L. 1980. Hydrology Comes of Age: Impact of the International Hydrological Decade. *EOS* 61 (53): 1241–1242.

OECD [Organization for Economic Co-operation and Development]. 1997. *Participatory Development and Good Governance.* Development Co-Operation Guideline Series, 1–34.

Pahl-Wostl, Claudia. 2007. Transitions Towards Adaptive Management of Water Facing Climate and Global Change. *Water Resources Management* 21: 49–62.

Peet, Richard, and Michael Watts (eds.). 2004. *Liberation Ecologies: Environment, Development, Social Movements.* New York: Routledge.

Pittock, Jamie, Karen Hussey, and Stephen Dovers (eds.). 2015. *Climate, Energy and Water.* Cambridge: Cambridge University Press.

Plumwood, Val. 2002. *Environmental Culture: The Ecological Crisis of Rationality.* New York: Routledge.

Postel, Sandra, and Brian Richter. 2003. *Rivers for Life: Managing Water for People and Nature.* Washington, DC: Island Press.

Postel, Sandra, Gretchen Daily, and Paul Ehrlich. 1996. Human Appropriation of Renewable Fresh Water. *Science* 271: 785–788.

Ribot, J. 2004. *Waiting for Democracy: The Politics of Choice in Natural Resource Decentralization.* Washington D.C.: World Resources Institute.

Ringler, Claudia, Anik Bhaduri, and Richard Lawford. 2013. The Nexus Across Water, Energy, Land and Food (Welf): Potential for Improved Resource Use Efficiency? *Current Opinion in Environmental Sustainability* 5 (6): 617–624.

Rockström, Johan, M. Falkenmark, J.A. (Tony) Allan, C. Folke, L. Gordon, A. Jägerskog, M. Kummu, M. Lannerstad, M. Meybeck, D. Molden, S. Postel, H.H.G. Savenije, U. Svedin, A. Turton, and O. Varis. 2014. The Unfolding Water Drama in the Anthropocene: Towards a Resilience-Based Perspective on Water for Global Sustainability. *Ecohydrology* 7: 1249–1261.

Rogers, Peter, and Alan W. Hall. 2003. Effective Water Governance. TAC Background Papers No. 7, Evander Novum, Sweden.

Russi, Daniela, Patrick ten Brink, Andrew Farmer, T. Badura, D. Coates, J. Förster, R. Kumar, and N. Davidson. 2013. *The Economics of Ecosystems and Biodiversity for Water and Wetlands.* London: IEEP/Ramsar Secretariat.

Sachs, Wolfgang. 1999. *Planet Dialectics: Explorations in Environment and Development.* New York: Zed Books.

Savenije, H.H.G., A.Y. Hoekstra, and P. van der Zaag. 2014. Evolving Water Science in the Anthropocene. *Hydrology and Earth System Sciences* 18: 319–332.

Schlosberg, David. 2004. Reconceiving Environmental Justice: Global Movements and Political Theories. *Environmental Politics* 13 (3): 517–540.

Schlosberg, David. 2010. Indigenous Struggles, Environmental Justice, and Community Capabilities. *Global Environmental Politics* 10 (4): 12–35.

Schmidt, Jeremy J. 2014. Historicising the Hydrosocial Cycle. *Water Alternatives* 7 (1): 220–234.

Schmidt, Jeremy J. 2017a. *Water: Abundance, Scarcity, and Security in the Age of Humanity.* New York: New York University Press.

Schmidt, Jeremy J. 2017b. Social Learning in the Anthropocene: Novel Challenges, Shadow Networks, and Ethical Practices. *Journal of Environmental Management* 193: 373–380.

Schmidt, Jeremy J., Peter G. Brown, and Christopher J. Orr. 2016. Ethics in the Anthropocene: A Research Agenda. *The Anthropocene Review* 3 (3): 188–200.

Scott, Colin. 1996. Science for the West, Myth for the Rest? The Case of James Bay Cree Knowledge Production. In *Naked Science: Anthropological Inquiry Into Boundaries, Power and Knowledge*, ed. L. Nader, 69–86. New York: Routledge.

Sedlak, David. 2014. *Water 4.0: The Past, Present and Future of the World's Most Vital Resource.* New Haven: Yale University Press.

Sneddon, Christopher. 2015. *Concrete Revolution: Large Dams, Cold War Geopolitics, and the U.S. Bureau of Reclamation.* Chicago: University of Chicago Press.

Snyder, Shane A., Paul Westerhoff, Yeomin Yoon, and David Sedlak. 2004. Pharmaceuticals, Personal Care Products, and Endocrine Disruptors in Water: Implications for the Water Industry. *Environmental Engineering Science* 20 (5): 449–469.

Suhardiman, Diana, Mark Giordano, and Francois Molle. 2012. Scalar Disconnect: The Logic of Transboundary Water Governance in the Mekong. *Society & Natural Resources* 25 (6): 572–586.

Sundberg, Juanita. 1998. NGO Landscapes in the Maya Biosphere Reserve, Guatemala. *Geographical Review* 88 (3): 388–412.

UNESCO. 2006. *Water: A Shared Responsibility.* Vol. 2 World Water Assessment Programme (United Nations). Oxford: Berghahn Books.

Vogel, David. 1997. Trading Up and Governing Across: Transnational Governance and Environmental Protection. *Journal of European Public Policy* 4 (4): 556–571.

Vörösmarty, Charles J., D. Lettenmaier, C. Lévqêue, M. Meybeck, C. Pahl-Wostl, J. Alcamo, W. Cosgrove, H. Grassl, H. Hoff, P. Kabat, F. Lansigan, R. Lawford, and R. Naiman. 2004. Humans Transforming the Global Water System. *EOS* 85: 513–516.

Vörösmarty, Charles J., P.B. McIntyre, M.O. Gessner, D. Dudgeon, A. Prusevich, P. Green, S. Glidden, S.E. Bunn, C.A. Sullivan, C. Reidy Liermann, and P.M. Davies. 2010. Global Threats to Human Water Security and River Biodiversity. *Nature* 467: 555–561.

Vörösmarty, Charles J., Charles Pahl-Wostl, Stuart E. Bunn, and RIchard Lawford. 2013. Global Water, the Anthropocene and the Transformation of a Science. *Current Opinion in Environmental Sustainability* 5 (6): 539–550.

Wang, Feng, Haitao Yin, and Shoude Li. 2010. China's Renewable Energy Policy: Commitments and Challenges. *Energy Policy* 38 (4): 1872–1878.

Warner, J., P. Wester, and A. Boldin. 2008. Going With the Flow: River Basins as the Natural Units for Water Management? *Water Policy* 10 (2): 121–138.

White, Gilbert F. 1978. Resources and Needs: Assessment of the World Water Situation. In *Water Development and Management: Proceedings of the United Nations Water Conference*, vol. 1, ed. A.K. Biswas, 1–46. New York: Pergamon Press.

White, Gilbert F., E. de Vries, H. unkerley, and J. Krutilla. 1962. *Economic and Social Aspects of Lower Mekong Development.* Bankgkok: Committee for Co-ordination of Investigations of the Lower Mekong Basin.

White Jr., Lynn. 1967. The Historical Roots of Our Ecological Crisis. *Science* 155: 1203–1207.

World Commission on Environment and Development. 1987. *Our Common Future.* Oxford: Oxford University Press.

Zimring, Carl A. 2016. *Clean and White: A History of Environmental Racism in the United States.* New York: NYU Press.

CHAPTER 3

Economies

Abstract This chapter examines how human economies have been understood in and through the practices informing global water governance, with a focus on the notion of economic value that shaped late twentieth-century sustainable development. It emphasizes the overlapping, partial, and abandoned economic techniques that have shaped global water governance and which subsequently complicate attempts at uniform or universal solutions to global challenges. It considers the progression of prominent economic ideas that have shaped global water governance: cost-benefit analysis, development economics and the Washington Consensus, water markets, common-pool resources, and ecosystem services valuation. It concludes by considering how the problems with securing water to a uniform economic system are increasingly being addressed by seeking nonuniform ways to secure water to the global economy through financial instruments that govern the supply chains of the water–energy–food–climate nexus.

Keywords Economics · Cost-benefit analysis · Common-pool resources Ecosystem services · Water–energy–food–climate nexus · Commons Privatization

Water has always been central to human economies, and to the vast diversity of formal and informal ways that human economies may be structured and function—from street vendors selling water by the

© The Author(s) 2017
J.J. Schmidt and N. Matthews, *Global Challenges in Water Governance*,
Global Challenges in Water Governance, DOI 10.1007/978-3-319-61503-5_3

jug, to markets for buying and selling rights to water for irrigation, to "sachet water" sold in sealed plastic bags, to local governments selling water to multinational corporations for production and sale of bottled water (Bakker 2010; Boelens et al. 2010; Gleick 2010; Stoler 2017). Likewise, numerous disciplines study the ways in which human economies value water in relation to individual, social, and environmental goods and within different political and institutional contexts. Given the diversity of human economies, and differences over how to conceptualize, manage, study, and ultimately govern water with respect to them, it is perhaps unsurprising that economic aspects of global governance have produced considerable contests. Many of these contests are not over whether water is valuable to human economies per se. Rather, they pivot around what is to count as "economic," how political and institutional structures are established and operate, and how techniques of calculating, distributing, and allocating water with respect to human economies are governed. Where there are significant differences, or where normative claims regarding what kind of economy is desirable are not aligned with the social contexts to which they are applied, open conflict can and has ensued.

Political or institutional structures that are misaligned with the social contexts of human economies they seek to govern can produce challenges for global water governance in at least two ways: First, what counts as "economic" might exclude existing ways water is valued either formally or informally. For example, without seeing market exchanges or an informal equivalent (e.g., the reciprocal sharing of labor), one might assume that, however water is functioning within a given social context, it isn't "economic" because it doesn't meet certain criteria; maybe there are no buyers or sellers of water, or perhaps water's economic value cannot be distinguished clearly from other activities that accompany it (Trawick 2010). For instance, some water economies might rely on the formal exchange of currency or labor only in a limited range of cases that are themselves determined by other social norms or gifts. Alternately, families in impoverished urban areas may share water connections in ways not easily accounted for in surveys or official accounts but which are nevertheless critical to household and informal economies (Anand 2017). By contrast, sometimes it may be middle- or upper-class citizens who push for water reforms to stop politicians from using water to manipulate the urban poor (Herrera 2017). In many cases, the changes wrought in the landscape to facilitate the use of water for particular economic orders, such as those of industrial production, are so thoroughly inculcated in particular forms of living and working that it appears natural for water to support particular political economies (Barca 2010). Across these cases, identifying or establishing what counts as economic, and how this counting is undertaken, requires deep contextual knowledge of a whole suite of social relations, some of which may not be accurately—or ethically—captured if what is to count as economic is predetermined (cf. Ballestero 2015).

Second, what counts as "economic" may capture only a small part of existing economic values and, by seeking to amplify them, have unintended effects. For example, in the nineteenth century, the priority of water rights in the western USA was based on when water was first put to beneficial use under a principle known as prior appropriation. The principle was specifically designed to counter forms of economic speculation by preventing wealthy firms from buying up water rights in ways that would undermine the vision of an agrarian society of American settlement (Schorr 2012). By tying water rights to actual water use, the accumulation of water rights by wealthy firms was rejected in favor of ensuring water supported local agrarian economies. In the 1980s, however, attempts were made to convert historical water rights into something more akin to private property. This amplified the individual rights held through the principle of prior appropriation in the name of enhancing the economic value of water. But this conversion captured only a small part of the water economies—both agrarian and municipal—that had by then come to rely on the upstream–downstream relationships of agrarian economies as a whole. As a result, amplifying individual ownership often did not fully consider how water rights, although held individually, were connected to broader community goods that may be negatively affected (even if unintentionally) when water rights were converted to private property in ways that changed the timing, efficiency, or location of water uses (Sax 1994; Freyfogle 1996). It also ignored how water in situ (i.e., in place) is already being used by ecological communities and conferring benefits to both humans and the environment (Wilkinson 1989). Rather than consider the broad suite of values existing arrangements supported (or ignored), it was assumed that the correlation between enhancing individual benefits and increased public well-being would hold in this case as it does in theories of neo-classical economics.

This chapter examines several of the main contours affecting how human economies have been understood in and through the practices informing global water governance. As the introductory chapter noted, the Dublin Principles emphasized water's economic value as part of ensuring it would not be overlooked in the sustainable development agenda that was crystallizing in the 1990s. As water governance rose in prominence through the late twentieth and early twenty-first centuries, its economic value continued to be an important facet of sustainability. The structure of this chapter roughly parallels the historical developments in water and economics that informed the transition from water management—the sites of actual decisions affecting water use—to water governance as sustainable development was articulated. The parallel is "rough" in the sense that the variability in time and over space regarding water and human economies means that there is no simple or shared trajectory, or even a set of predictable steps, to how water has been valued in economic terms. The parallel is also rough because differences in human economies have led to numerous contests, especially in attempts to put in place political or institutional structures that would provide uniformity between water governance and a liberalized global economy. In practice, many overlapping, partial, and abandoned economic techniques have shaped, and continue to shape, the contexts of global water governance. As such, this chapter does not attempt to give an exhaustive account of water in human economies—in fact, in its focus on the driving ideas that have linked water, sustainability, and governance, it reveals how narrow determinations of what counts as "economic" have often become and the challenges this narrowing gives rise to. With these caveats in mind, it considers the progression of several economic ideas that have shaped global water governance: (1) cost-benefit analysis, (2) development economics and water markets, (3) common-pool resources, and (4) ecosystem services valuation. It concludes by considering how the problems with securing water to a uniform economic system are increasingly being addressed by (5) seeking non-uniform ways to secure water to the global economy through financial instruments that govern the supply chains of the water–energy–food–climate nexus.

COST-BENEFIT ANALYSIS

Cost-benefit analysis, or CBA, emerged in the USA in the early twentieth century and was made a legal requirement for water planners in the US Army Corps of Engineers in the 1936 *Flood Control Act*. Cost-benefit analysis was part of an attempt to include a broader set of considerations, across both private and public forms of accounting, into the calculations used to determine whether and how water projects should proceed. There were two primary reasons for doing so: The first was the devastating impact of floods on lives, livelihoods, and both private and public property in the late 1920s and early 1930s. The second was to avoid partisanship by providing decision makers with objective and comprehensive ways to account for the merits of different courses of action. In its initial, and now famous formulation, action for flood control was to be justified "if the benefits to whomsoever they may accrue are in excess of the estimated costs, and if the lives and social security of people are otherwise adversely affected" (United States 1936, p. 1570).

In context, the adoption of cost-benefit analysis and its formulation in law was also designed to bring water under a more comprehensive form of governance. Since the late nineteenth century, US water planners had been concerned with how multiple uses of water by private individuals and firms may lead each to do what was best for their own interest, but that this would not lead to the maximum good for the national interest. During the presidency of Theodore Roosevelt in the early twentieth century, senior advisors on issues of resource conservation argued that the government must steward water as a public good in order to maximize the national interest. There was no agreement on how this should be done, but one of the key architects of conservation in the USA went so far as to argue that, because water was absolutely essential for the economy, the gold standard for currency should be replaced by a water standard based on the nation's hydrological endowment (Schmidt and Shrubsole 2013). That of course did not happen, yet it reflects the priority that has long been assigned to water's economic value and the need to maximize both private opportunity and public welfare in liberal systems of governance. Instead of using water as the material basis for currency, the USA formally legislated cost-benefit analysis in 1936, only 3 years after it had created what became one of the most influential models for comprehensive water management in the modern era—the Tennessee Valley Authority.

The goal of comprehensively managing water through the TVA shaped the institutional context for the objective techniques of cost-benefit analysis by requiring decision makers to bring together very different kinds of concerns under a common set of indices. Flood control was one concern, but so too were electricity, agriculture, industrial manufacturing, national security (the TVA became a key supplier of energy for wartime efforts), and urban demands. Added to considerations of economic impacts were multiple social concerns, from the private rights of individuals to those of municipalities, states, and federal agencies. In this context, the promise of cost-benefit analysis was that it would provide a method for comparing multiple factors based on a common, objective measure. Objectivity, in this context, was to be derived based on the conversion of costs and benefits into dollars and cents. This was designed to stem the subjective nature of water politics in which different constituencies and special interest groups captured the attention of decision makers. A common measure of comparison for costs and benefits, it was hoped, would provide for analysis in a manner free from subjective political, moral, or social biases. In short, cost-benefit analysis was "introduced to promote procedural regularity and to give public evidence of fairness in the selection of water projects" (Porter 1995, p. 149). Thus, part of the original rationale for cost-benefit analysis was to find a way to manage the differences in how water was related to multiple aspects of human economies without necessarily changing the political or institutional structures governing decisions on irrigation, municipal systems, or flood control. That is, cost-benefit analysis was a technical solution to a governance challenge.

From the 1950s to the 1970s, cost-benefit analysis grew in both sophistication and scope; especially influential during this period was the work that proceeded under the Harvard Water Program, which brought together policy makers, scientists, and economists to develop more comprehensive forms of estimating the impacts of water projects. Early computational models were designed and tested for multipurpose river basin projects—using multiple social and physical variables—with the goal of integrating multiple factors into a comprehensive plan that maximized net benefits (Reuss 2003). As the program matured, it incorporated both economic and noneconomic values into its analysis in order to provide as comprehensive picture as was possible to decision makers (Maass and Hufschmidt 1960). By the 1970s, water managers in the USA touted the advances in cost-benefit analysis as having fundamentally changed—for the better—the ways in which water decisions were taken. Gilbert White, for instance, believed this to be the case. Later, White (1978) took American ideas into the international arena by providing the first global assessment of water resources and needs at the 1977 UN Conference on Water in Mar del Plata.

Cost-benefit analysis, however, was never entirely objective or comprehensive. This became apparent when the international networks that arose after Mar del Plata tried to implement it. In the late 1970s and early 1980s, experts from France noted that cost-benefit analysis required judgments regarding which variables were to be used in calculating costs and benefits and also required judgments regarding the relative weights to assign chosen variables (Schmidt 2017). These practical judgments meant cost-benefit analysis was never value free. The need for judgments, however, also belied deeper problems with claims to objectivity. Douglas Kysar (2010) has argued that a basic deficit of cost-benefit analysis is that choosing to implement it is the outcome of political decisions that are shaped by the institutional structures, norms, and values of a given political community. If we lose sight of this, cost-benefit analysis has the appearance of being able to "regulate from nowhere" through claims to objectivity when in fact it is based on political and moral judgments; this is not only because a particular community chooses to deploy cost-benefit analysis in decision making but also because, in operation, cost-benefit analysis requires a whole suite of judgments, such as how to choose among different variables available for measure and how to weight chosen variables in determinations of costs or benefits. Finally, cost-benefit analysis can have perverse effects. For instance, mandating that benefits always exceed costs in water projects has led some water agencies to find ways to increase benefits in unanticipated ways. Draining wetlands so that land can be sold to developers (a benefit) has, for instance, been used as a way to justify flood control infrastructure projects (a cost) in ways that ignore the role of wetlands in reducing the extent and risks of floods (Kysar and McGarity 2006).

There were numerous challenges of meeting the demands for objectivity that cost-benefit analysis required. In addition to those above, variables not included in cost-benefit analysis were simply not assigned economic value. Despite these difficulties, cost-benefit analysis has been very influential in international development, where modernization projects throughout the latter half of the twentieth century often relied on it to determine the feasibility of different projects. Of course, costs and benefits of the project were not the only concerns. Political influence was another significant consideration as development programs played out in the context of the Cold War. These political dynamics further complicated the claims to objectivity in cost-benefit analysis. The effect was that cost-benefit analysis was often made instrumental to broader political interests, which was exactly opposite to why it was adopted in the first place.

DEVELOPMENT ECONOMICS AND WATER MARKETS

The use of cost-benefit analysis in international development throughout the twentieth century was also significant in shaping the program of global water governance that later emerged. Loans for large dams, investments in infrastructure for urban sanitation and drinking water, and improvements in irrigation technologies were not exclusively matters of dollars and cents. Instead, the ways in which costs and benefits fit with development programs were also matters of political economy. Decisions on which projects to fund, for instance, often involved a political calculus that reached well beyond the specifics of a particular water project. For instance, the Aswan High Dam in Egypt was a project the World Bank sought funding for from the USA, who ultimately decided against it for reasons that extended all the way to local politics—such as ensuring cotton grown by water from the new dam would not compete with that from producers in southern US states (Mitchell 2002; Schmidt 2017). The Aswan High Dam was, in turn, funded by the Soviet Union. Subsequently, the USA upped its support of water projects in Egypt substantially, not based on cost-benefit calculations exclusive to the benefits derived from water, but rather to ensure it did not lose influence in the region (Barnes 2014).

The political economy of development projects is not limited to questions regarding which countries gave loans to which projects or for what reasons. Rather, development economics are also important to understand how particular kinds of economic interventions affected both existing economies and the social relationships in which those economies operated. For instance, large-scale urban water development projects fostered new social understandings of water by creating new ways of thinking about public water supplies. As this took place, it affected understandings of what kinds of social and political relationships were appropriate for governing and managing the new forms of public water put in place by the development of urban supply systems (Bakker 2013). Elsewhere, such as in Iran, development programs for agricultural modernization sought to replace traditional farming techniques with new technologies and practices that would increase agricultural participation in the global economy. One way this took place was by increasing the number of pumps used to extract groundwater for cash crops. These shifts had significant consequences for Iran's existing agricultural economies and the long histories of social relationships, governance structures, and dispute resolution practices that "modern" development programs sought to replace (Foltz 2002; Balali et al. 2009). In particular, in the 1970s, at just the time when global water concerns were being recognized, "development" frequently meant increasing industrial productivity in agriculture, manufacturing, and other sectors. A common feature of development programs was an attempt to suspend political judgments about the desirability of different human economies through technical appeals to comprehensiveness or objective economic calculations (Li 2007). Despite this attempted suspension of judgments, however, it was evident that development was not simply a technical exercise in increasing economic benefits derived from water (or natural resources generally), but a deeply social and political transformation of diverse human economies into the more uniform relationships of production, scale, and labor demanded by industrial modernization (Ferguson 1990).

Development projects were not only pursued at national scales or constrained to large urban areas. They also frequently involved household and village-scale interventions where funding for various projects introduced new spheres of economic values and, at the same time, new legitimizing ideas for claiming rights to participate in those economic spheres (White et al. 1972). In global governance, value pluralism offers one way to understand how various kinds of rights are given legitimacy—from internationally recognized human rights, to the national rights guaranteed by constitutions, to rights associated with prevailing religions, customs, property regimes, or to the specific rights associated with particular development projects (Pradhan and Meinzen-Dick 2003). These different spheres of rights, and the values that support them, may conflict with one another. This can have significant impacts on the extent to which the social relationships of different economies incorporate or reject different development interventions. For example, establishing discrete, individual water rights can be problematic since, in many cases across Africa and Asia, local economies connect the access and use of land and water to one another and, in so doing, connect water not solely to an individual but also to social relationships among kin, households, and communities (Meinzen-Dick and Nkonya 2005). By the same token, recognizing that there are often many spheres of rights and values can be helpful in cases of oppression, such as when developmental, national, or international rights are employed to confront oppressive social relationships, such as those based on gender or race (Zwarteveen and Meinzen-Dick 2001).

By the 1980s, the complexity of water management increasingly led to the conclusion that large, top-down development agencies were ill equipped to maximize the benefits to be derived from water. There was, in short, a shift away from attempts to use a uniform set of economic techniques (i.e., cost-benefit analysis) to connect different sectors as well as a shift away from attempts to use uniform development programs to enhance economic productivity (i.e., state-planned industrialism). These large-scale efforts were seen as necessary, but not sufficient, due in some measure to the fact that single agencies and single programs did not have the knowledge required for the kinds of multipurpose water management challenges posed by industrializing societies. Partially in response to the economic ideas of Friedrich Hayek (1945), these limitations came to be seen as intrinsic to the complex nature of modern economies, so were not in principal resolvable by enhancing the technical or managerial powers of water management institutions (see Boelens et al. 2010). This led many to the conclusion that funds spent on institutions should be minimized in favor of increasing reliance on the transactions individuals or firms might make regarding water. In tandem with the changing logic of economic theories, the amount of development aid available for water significantly dropped off throughout the 1990s and private investment rose (Briscoe 1999). This shift, as we will see below, was a portent to major social and political transitions across both developed and developing economies due to the fact that water's functional integration in existing economies had often operated to serve the interests of the public good, as discussed above.

The international effects of these shifts in ideology, often referred to as the "Washington Consensus," were that developing countries were frequently disciplined into liberalizing their economies in ways that encouraged global integration with trade, capital, and labor markets, often with little consideration for the associated environmental impacts (Gore 2000). When it came to water, shifts from state-led programs to market-based institutions were complicated and uneven. Water itself was rarely traded in the global economy—it is too heavy to economically ship in most cases, for instance—so fitting water with liberalization policies required reordering the social, political, and legal values that would allow it to function economically in an era of globalization. International development agencies, such as the World Bank, took on an enhanced role in shaping economic development throughout the 1990s—at precisely the same time that IWRM was gaining international influence—and frequently tied development aid to demands for structural adjustments that premised project funding on revisions to laws affecting environmental regulation, private property rights, and the privatization of public water services (Goldman 2005).

This era of "green neoliberalism," as it is sometimes referred to, promoted economic reforms based on the premise that, where states agencies had failed, economic instruments could succeed in both enhancing economic development and meeting environmental goals. Karen Bakker (2014) identifies five dimensions to how, broadly speaking, liberalization policies affected water in the post-Washington Consensus era: (1) there has been an increased trend toward the private ownership of water by individuals or firms; (2) there has been an increase in the commercialization of existing water management institutions (i.e., public utilities); (3) there has been a rise in market or trading mechanisms for reallocating water to areas of higher economic value; (4) new techniques have been developed for valuing different aspects of water, such as water for the environment; and (5) there has been a deregulation and/or decentralization (predominantly from state agencies to private organizations) of the institutions that provide oversight on issues of regulatory compliance and environmental monitoring. Importantly, these different aspects of liberalization are not always separate from one another and neither do they operate in smooth or universal fashion. By exploring a case study from Chile, different liberalization policies can be seen for how they operated in the same place, and with somewhat different rationales, despite the uniform "consensus" claimed for market reforms, property rights, and the commercialization of public utilities.

In 1981, under the military rule of Augusto Pinochet, Chile adopted a new Water Code. The Water Code created strong private property rights in water with the expectation that a market for buying and selling clearly defined water rights would emerge. Initially, international agencies such as the World Bank lauded and encouraged the Chilean exercise. But after a decade, empirical work revealed the market remained inactive for several reasons: the geography of Chile made transferring water physically difficult; there was confusion and contest over water titles and historical records of ownership; both farmers and people in general were hesitant to treat water as a commodity; and finally, inconsistent prices worried those interested in using markets to acquire or sell water rights (Bauer 2004a, b). By 1990, a new government had been put in place and, again with encouragement from the World Bank, was aiming to privatize water supply and sanitation due to the large costs it faced for infrastructure by the mid-1990s. By 1999, Chile's public utilities had been privatized. The privately run utilities provide nearly universal coverage at competitive rates and stand out from Latin American countries, such as Bolivia, where privatization created intense political conflict. But why did the Chilean case succeed? As Madeleine Baer (2014) shows, Chile's privatization program did not succeed because the market fixed government inefficiencies, but because private companies took over model institutions that were already effectively suited to their social context.

The Chilean case exemplifies a whole spectrum of activity associated with economic shifts in the late twentieth century. Elsewhere, water markets emerged, failed, and were redesigned to be more effective in Australia, the western USA, China, Canada, and South Africa (Garrick 2015). Comparative analysis of water markets both across institutions and over time suggests that three key features are central to understanding how markets fit with integrated forms of water management and governance (see Grafton et al. 2011): (1) the institutional foundation of the market is key to ensuring that it functions with respect to concerns regarding factors like the recognition of public interests, administrative capacity, and instances of market failure; (2) the ability of the market to improve economic efficiency must be considered with respect to the size and extent of the market, the nature of water rights, and the availability of price information; and (3) the relationship between the market and environmental sustainability must ensure there is adequate scientific data, adequate provisions for environmental water flows, and linkages to issues of water quality and not only quantity. In short, after decades of experimenting with water markets, their use requires both initial and ongoing governance of knowledge, communication, and the actions of multiple actors across social and environmental domains.

The Chilean case also reveals some of the complexities surrounding the privatization of municipal utilities. Private water corporations have sometimes had success if institutions, infrastructure, and incentives align favorably. Often, however, this has not been the case, with the result that many water utilities were (or are) being returned to the control of municipal governments—a process known as re-municipalization (Hall et al. 2013). In some cases, both domestic and international policies have encouraged public water utilities to become profit-generating organizations, which has encouraged some large urban utilities to extend water services to new areas of demand that private companies may not be interested in pursuing given that these areas are often only economically serviceable because of the economies of scale of large utilities (Furlong 2016). The extension of urban water connections, however, nevertheless remains politically important as people use water to make claims on the state, often within broader political efforts that connect water, urban infrastructure, and citizenship. These challenges are of course made yet more complex by the dynamics of cities themselves, particularly the combination of politics with rapid urbanization, informal settlements, and demands for services (Anand 2017; Herrera 2017; Bjorkman 2015; Ioris 2015).

In broad terms, what might be taken from this discussion is that, in several important respects, the Washington Consensus failed to account for numerous social, political, and environmental factors that affect different human economies. As one expert put it, by the mid-1990s international water management was "a dialectic between two philosophical norms; one, the rational analytic model, often called the planning norm, and two, the utilitarian or free-market model, often couched in terms of privatization" (Priscoli 1996, p. 30). That is, the focus on managing water at the point of decision making—whether through government planning or economic exchanges—was like a pendulum swinging back and forth between states versus markets. Of especial importance was that the Washington Consensus also did not explicitly consider biophysical factors, such climate change, and assumed its view of economic efficiency would be environmentally superior to the alternatives (Dellapenna 2008). From an economic perspective, changes at the point of water use decisions—the traditional domain of water management—had to be complemented by a broader focus on water governance that would incorporate political, institutional, and environmental concerns into economic considerations.

COMMON-POOL RESOURCES: BEYOND STATES VERSUS MARKETS

Economic conceptions of water extend far beyond matters of marginal pricing, markets for exchange, or the total utility government planners might estimate from different water projects (Hanemann 2006). Often, the distinction between states and markets conceals more than it reveals about the complex ways in which human economies operate. In an effort to move beyond "states versus markets," a number of scholars have focused on the institutions that allow for collective governance of water to succeed through the different forms that organizations, norms, or practices may take. The key starting point for these theorists is that water governance (and environmental governance generally) avoid the false dichotomy in which humans inevitably exhaust commonly held goods and therefore must either be governed strictly by states to limit consumption or through market mechanisms that will price resources sufficiently to conserve them in the long run. This false dichotomy, according to the leading intellectuals of common-pool resource theory like the late Elinor Ostrom (1990), obscures the fact that many human economies have flourished through ways of sharing water that are not simply combinations of states versus markets (see also Ostrom and Gardner 1993; Chambers 1988).

For common-pool resource scholars, the problems of the state/market dichotomy are often connected to Garrett Hardin's (1968) classic essay on "The Tragedy of the Commons." There, Hardin argued that, in a context where individuals share a common resource, each individual has a rational incentive to take more from the commons so as to benefit from it as much as possible. The result of every individual pursuing this strategy is the overuse, and ultimately tragic collapse, of the commons— such as the overpumping of groundwater that ultimately leaves the entire community bereft of its resource base. Hardin concluded that private enclosure of the commons was a preferred way to ensure individuals viewed the stewardship of resources as in their own resources. The alternative, he argued, was for there to be "mutual coercion mutually agreed upon." That is, for Hardin there were only two options: either private property and market prices or state control.

By contrast, common-pool resource scholars point out several errors in Hardin's account. The first is that many common goods do in fact have social rules and institutions that govern them. So the only cases where his thesis might reasonably apply are where there are "open-access" resources, such as in the limited number of cases where there is a true institutional vacuum. The second error of Hardin's account is that individuals are not locked into zero-sum games where individuals each pursue their own rational interest without respect for their collective effects. In fact, many communities not only have social norms that limit overall resource use, but many individuals frequently see that it is rational for them to limit their use of the commons to preserve the resource base that they rely on. Once these two errors are corrected, a third comes into view; the forced dichotomy that reduces solutions to resource sharing dilemmas to either state coercion or private property—or a combination of the two—does not adequately account for the diversity or complexity of human economies (Ostrom 2010a, b).

From the perspective of common-pool resource theory, there are no institutional arrangements that can provide universal templates for governing water resources (Ostrom 2007). Rather, the complexity of institutions required for water governance crisscrosses politics, law, custom, gender, religion, and values with the upshot that there are no panaceas (Meinzen-Dick 2007). Further confounding attempts to establish one-size-fits-all arrangements are two key considerations. The first is that existing practices for collectively governing shared resources have in some cases evolved for centuries (or longer) and include a combination of codified rules and conventional, learned practices that follow the contours of collective responses to advances in technology, environmental changes, as well as to responses to forces of colonialism, modernization, and globalization. The second is that the close coupling of water systems and practices of collective action has fostered unique coevolutionary relationships between social and ecological communities at multiple scales as communities adapt to changes in their environmental and social conditions (Ostrom 2009). Taken together, it is unlikely that global water governance will succeed in contexts of common-pool resource sharing based on establishing ideal institutional types, such as states or markets, since these cannot adequately account for social and ecological diversity, complexity, or change. Further, the reduction in complexity required to apply ideal types may reduce the range and flexibility of common-pool arrangements, which will likely reduce their resilience.

Common-pool resource theorists do not see complexity as a challenge to be managed. Rather, complexity is a source of economic and institutional advantage, particularly because under conditions of global environmental change complexity often provides greater redundancy for dealing with disturbances. As such, common-pool resource theory considers how economic goods emerge as an outcome of two functions (Ostrom 2003): The first is a production function in which goods are produced through collective action. For instance, it is through collective action that irrigation infrastructure networks are often built and maintained as individuals work together to build and maintain conveyance structures. The second function is one of allocation, where institutional norms and rules legitimate how individuals holding entitlements share the goods produced through collective action. Sometimes, the two functions are related, such as when participating in the maintenance of an irrigation system is part of securing an allocation of water. But this is not necessarily the case as other social relationships (e.g., kinship) might bear on who has an entitlement to goods. Importantly, at scales from local irrigation systems to large water-trading schemes, the attention of common-pool resource approaches is focused on the selection and use of institutional mechanisms—policies, markets, or practices—that fit the context (Garrick 2015).

Despite their attention to the complexity of human resource economies, common-pool resource approaches nevertheless face difficulties. One is that they tend to treat nature as inherently passive—economic goods can be produced and allocated as discrete "resource units" without much consideration for alternative ways of thinking about human relationships with the natural world. This sits uneasily with many indigenous practices in which non-human agency is not a metaphor for ecological action, but an orientation to social relationships and governance (Schmidt and Dowsley 2010). A second is that the institutional focus of common-pool resource theory can come at the expense of understanding the unique moral economies that relationships to water are governed through. Social scientists have argued, for instance, that principles of equity, transparency, and fairness are what often make successful institutions work (Trawick et al. 2014). The moral economy of water, on this account, is not often treated in explanations of common-pool resource theories that instead appeal to (modified) game-theoretic understandings of rationality rather than ethnographic research regarding customary norms. A third is that complexity is often anathema to policy makers, particularly when certainty is valorized as a precondition for effective pricing mechanisms or for justifying the use of development funds. The default position of viewing complexity as a problem in development economics can be further compounded by demands that governance structures produce shareholder value (Cooper 2010). In such cases, the drive for certainty in the pursuit of one view of what counts as "economic" (i.e., shareholder value) can be significantly out of step with both formal and informal human economies that water supports. In response to these challenges, common-pool resource theorists have increasingly made connections to discourses on resilience that have gained considerable purchase in the arena of governance and sustainability and, related to this, forms of economic valuation that seek to provide certainty without unnecessarily reducing the complex ways in which multiple human economies value water. One of these has been through the valuation of water's ecosystem services.

ECOSYSTEM SERVICES

Classic and neo-classical models of economic valuation often do not adequately capture community-level values or the values that water provides as a condition for—not only as a component of—healthy economies (Russo and Smith 2013). As environmental concerns became increasingly pressing in the latter half of the twentieth century, new techniques of economic valuation were created in order to enhance sustainable governance. In 1997, Robert Costanza and his colleagues argued that the ecosystem services nature provides to humanity and the capital "stock" of nature from which these services "flow" average (conservatively) $33 trillion USD annually (Costanza et al. 1997). Although some economists criticized this and other attempts at ecosystem valuation for violating microeconomic principles of diminishing marginal utility, budget constraints, and comparison of most feasible alternatives (e.g., Pearce 1998; Bockstael et al. 2000), the work of Costanza and others is frequently cited as a powerful way to incorporate environmental concerns into existing paradigms of economic valuation.

Ecosystem services valuation gained prominence after featuring centrally in the Millennium Ecosystem Assessment that published its results in 2005. The Millennium Ecosystem Assessment brought together thousands of scientists from around the globe to give an empirical picture of ecosystems, their interactions, and values, at scales from the local to the global. In its simplest form, ecosystem services valuation asks: What would it cost to replace through technology or human activity the services that nature currently provides for free? For instance, how much would it cost to treat drinking water if a wetland that currently does much of this for free is filled in for a new airport? By taking this line of questioning and developing metrics for assessing the value of ecosystem services, the goal is to assess what kinds of trade-offs exist in development scenarios that formerly assigned little or no value to ecosystems. In a manner not too dissimilar to cost-benefit analysis, one of the benefits of ecosystem services valuation is that it provides decision makers with a basis for comparing different governance scenarios across a suite of different domains—from environmental sustainability to local livelihood constraints. Of course, ecosystem services do not exist only with respect to water, and ecosystem services valuation has been applied to a range of considerations, including forestry and biodiversity conservation (Dempsey 2016).

The intuitive question driving ecosystem services valuation runs almost immediately into the complexity of finding an answer. Here, we present some of the initial difficulties, followed by lessons learned. As with other techniques of economic valuation, such as cost-benefit analysis, ecosystem services valuation requires a whole series of judgments about which services to measure, how to measure them, and what the likely effects of different trade-offs might be. In the USA, for instance, programs of wetland banking reveal the complexity of finding uniform prices given that wetland environments are highly variable (Robertson 2007). In these cases, companies or individuals that create environmental damages in one area can pay to maintain ecosystem services elsewhere, while other firms or individuals can preserve or create wetlands so that they have "credits" to sell. Similar programs also exist for river restoration and both face numerous difficulties: One is that the complexity of aquatic ecology does not lend itself to establishing uniform areas of "credit" when development "debits" the environment somewhere else. A second is that these schemes rely on ecosystem services as fitting a particular kind of economy shaped by regulatory rules that may curtail the ability of these schemes to use science as the basis for comparisons of economic or ecological value (Lave 2012; Robertson 2012). A third is that there can be many different ways of valuing nature, and its contributions to the quality of life may vary culturally or institutionally in ways that are incommensurable and so not reducible to a single yardstick (Martinez-Alier et al. 1998). Finally, ecosystem services valuation faces the general problem that the types of values it identifies are limited to those that can be valued by particular economic techniques. Seeing ecosystems as capital "stock" from which services "flow" has excludes many valuable ecological functions—and the sciences through which they are understood—that don't mirror the stock-flow models of environmental economics (Norgaard 2010).

Despite difficulties with ecosystem services valuation, programs that have combined payments for ecosystem services with programs of poverty reduction, such as in South Africa, have had success in both improving local livelihoods by providing wage labor opportunities and improving watershed environments upstream of those paying for water's ecosystem services (Turpie et al. 2008). Elsewhere, such as in upland regions of the Andes, efforts to implement payments for ecosystem services have reinforced the important role of governance, particularly with respect to ensuring that: (1) payments for ecosystem services are proportional to the values being sought; (2) the spatial areas required for improving or maintaining ecosystem services are targeted correctly; and (3) that there are adequate financial arrangements in place to keep programs focused on the key goals and areas of economic exchange (Goldman-Benner et al. 2012).

The learning curve associated with payments for ecosystem services, however, has not been simply a matter of matching scientific assessments to economic development programs. In fact, across several Andean communities involved in payments for ecosystem services schemes, the introduction of new economic relationships for communities who provide the "services" for downstream "consumers" of a healthy environment has not adequately dealt with the complex social relationships that connect land and water rights to social practices or existing structures of governance (Boelens et al. 2014). These complexities are related to many factors, from customary water uses and their relation to social relations and land tenure, to the effects of previous development programs, social inequality, to historical and ongoing injustices that have arisen from water use in other sectors, such as mining. In these complex social, historical, and political contexts, the attempt to parse payments for ecosystem services from other factors affecting water use is often seen as another form of commodification that fails to appreciate the depth of social relationships affecting issues of access and fairness that are central to governance concerns (Kosoy and Corbera 2010).

The Water–Energy–Food–Climate Nexus

Adding to the numerous programs of water markets, privatization, and payments for ecosystem services that have been experimented with in the past several decades, global development agencies are increasingly seeking to meet development challenges through more flexible arrangements. Principal among these has been the introduction of the water–energy–food–climate nexus (hereafter: nexus) as a concept that can both retain the promise of holistic water governance and, instead of seeking integration with a single institutional model, achieve sustainability through the integrated management of supply chains that affect the production of water, food, and energy (Schmidt 2017). Since it was established in global sustainability discourse, the nexus has rapidly risen as a way to keep the economic value of water central without separating its value from the numerous ways that water functions in all kinds of human societies. As the World Economic Forum (2011) argued, many of the global challenges regarding energy, food, and climate security are intricately linked with the value—or lack of value—assigned to water.

A key starting point for the nexus is that water is already connected in multiple ways to food and energy production and the supply chains that deliver goods to consumers. That is, water already facilitates production, manufacturing, use, and disposal of vast portions both the material and energy required by human economies. From the perspective of governance, this nexus of concerns does not require a single institutional format, but it does require a set of institutions that ensure water's value is recognized in any given institutional structure. In this sense, water's economic valuation in the nexus is explicitly a governance challenge, wherein water's economic value has moved from points of decision making—which are too numerous to manage—to a focus on the structures through which decisions about value are made. These concerns were given special emphasis at a conference in Bonn that took place ahead of the Rio+20 Conference scheduled for 2012. At Bonn, water's key role in meeting sustainability challenges was tied to the growing demands for water, energy, and food in a context where urbanized, global economies faced distinct challenges from climate change (Hoff 2011).

In principle, effectively governing the nexus will result in both improved efficiency in water use and improved environmental outcomes as water is conserved (Ringler et al. 2013). Similarly, the flexibility made possible by connecting sustainability to the nexus provides for the kind of adaptive, institutional resilience that a focus on states versus markets could not provide (Hussey and Pittock 2012; Scott et al. 2015). By using supply chains as the scale for economic valuation of the nexus, the complex ways in which supply chains crisscross over national and international boundaries provide a window into where, and through what mechanisms, water is being undervalued. This, in turn, provides points of incision for appropriately valuing water by identifying where it is structurally undervalued and establishing institutions that reveal water's value in those contexts. In addition to providing a novel set of terms for understanding risks to water security, the nexus has also been promoted by international agencies, such as UN Water, as key to international development given the intersecting challenges posed by energy, food, and climate crises (UN Water 2014).

Although the nexus is increasingly used in discourses on the Sustainable Development Goals, and in discussions of how best to finance infrastructure needs in a variety of development contexts, many scholars have identified difficulties that need to be addressed. One is that a focus on supply chain management does not ensure that ecosystem needs at local or regional levels will be sustainably governed (Allan and Matthews 2016; Lawford et al. 2013). For instance, it may be economically defensible to heavily withdraw or contaminate water in a local area if the overall value of the supply chain is deemed superior to the alternatives. Or the overall health of a local watershed may simply not be of concern if supply chain needs do not value holistic relationships between water and other biophysical and social relationships. This, of course, is an objectionable outcome if the point of the nexus is to ensure sustainability. In this regard, the trade-offs implied by the Sustainable Development Goals, such as between agriculture, nutrition, and energy provision, are not always clearly calculated with respect to their cumulative demands on limited water supplies or their economic implications (Ringler et al. 2016). As noted in Chap. 2, some countries may prioritize the social benefits of developing food or energy over environmental benefits.

A related difficulty is that actors and procedures making decisions about supply chain management are not typically either democratically constrained or inclusive of those most negatively affected by environmental injustice. Rather, elite, undemocratic networks seeking to maximize particular forms of economic value are typically those making decisions regarding supply chain management (Downey 2015). This means that governing water through appeals to the nexus will need to significantly reform either how democratic decisions are conceived of, or address the unequal power relations affecting how supply chain decisions are made. A final difficulty is that the supply chain focus of the nexus has not yet carefully considered how the sustainability of local livelihoods may be affected if supply chain security gains prominence over social considerations of development (Biggs et al. 2015). That is, if we ask, "Who is development for?" then concerns over how an emphasis on supply chains supports social, democratic aims of development must figure centrally in the response.

Conclusion

As water governance has ascended, the value of water to human economies has figured centrally. The overall tenor of these efforts—from the Dublin Principles onward—has been to use economic tools to help coordinate the tasks of governance. Frequently, however, these economic tools, from cost-benefit analysis through to markets, privatization, and ecosystem services valuation, have suffered the deficit of applying economic ideals to particular contexts. In a manner not dissimilar to the reflection of western values in the turn to governance itself, the result is a development path that repeats the patterns and political economy of globalization without learning lessons from the failures of what Ken Conca (2006) calls the twentieth-century era of "pushing rivers around." At present, for instance, some 3700 major dams are under construction or slated for development, which will increase hydroelectricity production 73% globally but lead to the challenges for biodiversity and local livelihoods discussed last chapter (Zarfl et al. 2015). These developments will further impact the steady loss of ecosystem services, which Costanza et al. (2014) recently estimated to be declining anywhere from $4 to 20 trillion USD per year due to land-cover change. In this sense, even if disagreements remain regarding the precise techniques of measurement and valuation, there is a much broader consensus that anthropogenic drivers of environmental transformation have broad social and environmental impacts that are both extensive and difficult to predict in terms of their full consequences.

Attempts to rearrange governance structures to promote economic ideals through reforms to property rights, public services, and environmental laws have often been successful only in a limited range of cases, while their failures have had social effects at scales from mundane, everyday practices through to moments of political upheaval. The alternative to applying top-down ideas about what "counts" in terms of economics would be to identify how human economies already value water through a variety of formal and informal relationships. Identifying the moral economies that attend numerous, successful forms of water sharing is crucial to understanding why governance reforms succeed or fail as individuals, groups, and communities practice different ways of sharing water. These broader social considerations are increasingly recognized as keys to sustainability, such as in the Sustainable Development Goals, where considerations of inequality are not limited solely to the regions of the Global South typically targeted for "development" programs but extend more broadly to any context where inequalities persist. In this sense, one way to interpret the normative call of the Sustainable Development Goals is not only environmentally—where the economy exists as a subsystem of the Earth system—but also socially, where the economy exists as one of the several ways in which water is valued by societies.

REFERENCES

Allan, J.A. (Tony), and Nate Matthews. 2016. The Water, Energy, and Food Nexus and Ecosystems: The Political Economy of Food and Non-food Supply Chains. In *The Water, Food, Energy, and Climate Nexus: Challenges and an Agenda for Action*, ed. Felix Dodds and Jamie Bartram, 78–89. London: Routledge.

Anand, Nikhil. 2017. *Hydraulic City: Water and the Infrastructures of Citizenship in Mumbai*. Durham: Duke University Press.

Baer, Madeline. 2014. Private Water, Public Good: Water Privatization and State Capacity in Chile. *Studies in Comparative International Development* 49 (2): 141–167.

Bakker, Karen. 2010. *Privatizing Water: Governance Failure and the World's Urban Water Crisis*. Ithaca: Cornell University Press.

Ballestero, Andrea. 2015. The Ethics of a Formula: Calculating a Financial-Humanitarian Price for Water. *American Ethnologist* 42 (2): 262–278.

Barca, Stefania. 2010. *Enclosing Water: Nature and Political Economy in a Mediterranean Valley 1796–1916*. Isle of Harris: White Horse Press.

Boelens, Rutgerd, D. Getches, and A. Guerva-Gill (eds.). 2010. *Out of the Mainstream: Water Rights, Politics and Identity*. London: Earthscan.

Bakker, Karen. 2013. Constructing 'Public' Water: The World Bank, Urban Water Supply, and the Biopolitics of Development. *Environment and Planning D: Society and Space* 31 (2): 280–300.

Bakker, Karen. 2014. The Business of Water: Market Environmentalism in the Water Sector. *Annual Review of Environment and Resources* 39: 469–494.

Balali, M.R., J. Keulartz, and M. Korthals. 2009. Reflexive Water Management in Arid Regions: The Case of Iran. *Environmental Values* 18: 91–112.

Barnes, Jessica. 2014. *Cultivating the Nile: Everyday Politics of Water in Egypt*. Durham: Duke University Press.

Bauer, Carl. 2004a. Results of Chilean Water Markets: Empirical Research Since 1990. *Water Resources Research* 40 (9): W09S06.

Bauer, Carl. 2004b. *Siren Song: Chilean Water Law as a Model for International Reform*. Washington, DC: Resources for the Future.

Biggs, Eloise M., Eleanor Bruce, Bryan Boruff, John Duncan, Julia Horsley, Natasha Pauli, Kellie McNeill, Andreas Neef, Floris Van Ogtrop, Jayne Curnow, Billy Haworth, Stephanie Duce, and Yukihiro Imanari. 2015. Sustainable Development and the Water-Energy-food Nexus: A Perspective on Livelihoods. *Environmental Science & Policy* 54: 389–397.

Björkman, Lisa. 2015. *Pipe Politics, Contested Waters: Embedded Infrastructures of Millennial Mumbai*. Durham: Duke University Press.

Bockstael, Nancy, A. Myrick Freeman, Raymond Kopp, Paul Portney, and Kerry Smith. 2000. On Measuring Economic Values for Nature. *Environmental Science and Technology* 34 (8): 1384–1389.

Boelens, Rutgerd, Jaime Hoogesteger, and Jean Carlo Rodriguez de Francisco. 2014. Commoditizing Water Territories: The Clash Between Andean Water Rights Cultures and Payment for Environmental Services Policies. *Capitalism Nature Socialism* 25 (3): 84–102.

Briscoe, John. 1999. The Changing Face of Water Infrastructure Financing in Developing Countries. *International Journal of Water Resources Development* 15 (3): 301–308.

Chambers, R. 1988. *Managing Canal Irrigation: Practical Analysis From South Asia*. New York: Cambridge University Press.

Conca, Ken. 2006. *Governing Water: Contentious Transnational Politics and Global Institution Building*. Cambridge, MA: MIT Press.

Cooper, Melinda. 2010. Turbulent Worlds: Financial Markets and Environmental Crisis. *Theory, Culture & Society* 27 (2–3): 167–190.

Costanza, Robert, Ralph d'Arge, Rudolf de Groot, Stephen Farber, Monica Grasso, Bruce Hannon, Karin Limburg, Shahid Naeem, Robert V. O'Neill, Jose Paruelo, Robert G. Raskin, Paul Sutton, and Marhan van den Belt. 1997. The Value of the World's Ecosystem Services and Natural Capital. *Nature* 387: 253–260.

Costanza, Robert, Rudolf de Groot, Paul Sutton, Sander van der Ploeg, Sharolyn J. Anderson, Ida Kubiszewski, Stephen Farber, and R. Kerry Turner. 2014. Changes in the Global Value of Ecosystem Services. *Global Environmental Change* 26: 152–158.

Dellapenna, Joseph W. 2008. Climate Disruption, the Washington Consensus, and Water Law Reform. *Temple Law Review* 81: 383–432.

Dempsey, Jessica. 2016. *Enterprising Nature: Economics, Markets, and Finance in Global Biodiversity Politics*. Chichester: Wiley.

Downey, Liam. 2015. *Inequality, Democracy, and the Environment*. New York: New York University Press.

Ferguson, James. 1990. *The Anti-Politics Machine: "Development", Depoliticization, and Bureaucratic Power in Lesotho*. Cambridge: Cambridge University Press.

Foltz, Richard C. 2002. Iran's Water Crisis: Cultural, Political, and Ethical Dimensions. *Journal of Agriculture and Environmental Ethics* 15: 357–380.

Freyfogle, Eric. 1996. Water Rights and the Common Wealth. *Environmental Law* 26: 27–51.

Furlong, Kathryn. 2016. *Leaky Governance: Alternative Service Delivery and the Myth of Water Utility Independence.* Vancouver: UBC Press.

Garrick, Dustin E. 2015. *Water Allocation in Rivers Under Pressure: Water Trading, Transaction Costs and Transboundary Governance in the Western US and Australia.* Williston: Edward Elgar Publishing.

Gleick, Peter H. 2010. *Bottled and Sold: The Story of Our Obsession With Bottled Water.* Washington, DC: Island Press.

Goldman, Michael. 2005. *Imperial Nature: The World Bank and Struggles for Justice in the Age of Globalization.* New Haven: Yale University Press.

Goldman-Benner, Rebecca L., Silvia Benitez, Timothy Boucher, Alejandro Calvache, Gretchen Daily, Peter Kareiva, Timm Kroeger, and Aurelio Ramos. 2012. Water Funds and Payments for Ecosystem Services: Practice Learns From Theory and Theory Can Learn From Practice. *Oryx* 46 (1): 55–63.

Gore, Charles. 2000. The Rise and Fall of the Washington Consensus as a Paradigm for Developing Countries. *World Development* 28 (5): 789–804.

Grafton, R. Quentin, Gary Libecap, Samuel McGlennon, Clay Landry, and Bob O'Brien. 2011. An Integrated Assessment of Water Markets: A Cross-Country Comparison. *Review of Environmental Economics and Policy* 5 (2): 219–239.

Hall, David, Emanuele Lobina, and Philipp Terhorst. 2013. Re-Municipalisation in the the Early Twenty-First Century: Water in France and Energy in Germany. *International Review of Applied Economics* 2: 193–214.

Hanemann, W.M. 2006. The Economic Conception of Water. In *Water Crisis: Myth or Reality?* ed. P.P. Rogers, M.R. Llamas, and L. Martinez-Cortina, 61–91. London: Taylor and Francis.

Hardin, Garrett. 1968. The Tragedy of the Commons. *Science* 162 (3859): 1243–1248.

Hayek, Friedrich A. 1945. The Use of Knowledge in Society. *The American Economic Review* 35 (4): 519–530.

Herrera, Veronica. 2017. *Water and Politics: Clientelism and Reform in Urban Mexico.* Ann Arbor: University of Michigan Press.

Hoff, Holger. 2011. *Understanding the Nexus. Background Paper for the Bonn 2011 Conference: The Water, Energy and Food Security Nexus.* Stockholm: Stockholm Environment Institute.

Hussey, Karen, and Jamie Pittock. 2012. The Energy-Water Nexus: Managing the Links Between Energy and Water for a Sustainable Future. *Ecology and Society* 17 (1): 31.

Ioris, Antonio. 2015. *Water, State and the City.* London: Palgrave.

Kosoy, Nicolás, and Esteve Corbera. 2010. Payments for Ecosystem Services as Commodity Fetishism. *Ecological Economics* 69: 1228–1236.

Kysar, Douglas. 2010. *Regulating From Nowhere: Environmental Law and the Search for Objectivity*. New Haven: Yale University Press.

Kysar, Douglas, and McGarity. 2006. Did NEPA Drown New Orleans? The Levees, the Blame Game, and the Hazards of Hindsight. *Duke Law Journal* 56: 179–235.

Lave, Rebecca. 2012. *Fields and Streams: Stream Restoration, Neoliberalism, and the Future of Environmental Science*. Athens: University of Georgia Press.

Lawford, Richard, Janos Bogardi, Sina Marx, Sharad Jain, Claudia Pahl-Wostl, Kathrin Knüppe, Claudia Ringler, Felino Lansigan, and Francisco Meza. 2013. Basin Perspectives on the Water-Energy-food Security Nexus. *Current Opinion in Environmental Sustainability* 5 (6): 607–16.

Li, Tania Murray. 2007. *The Will to Improve: Governmentality, Development, and the Practice of Politics*. Durham: Duke University Press.

Maass, Arthur, and Maynard M. Hufschmidt. 1960. Report on the Harvard Program of Research in Water Resources Development. In *Resources Development: Frontiers for Research*, ed. Franklin S. Pollak, 133–179. Boulder: University of Colorado Press.

Martinez-Alier, Joan, Giuseppe Munda, and John O'Neill. 1998. Weak Comparability of Values as a Foundation for Ecological Economics. *Ecological Economics* 26 (3): 277–286.

Meinzen-Dick, Ruth. 2007. Beyond Panaceas in Water Institutions. *Proceedings of the National Academy of Sciences* 104 (39): 15200–15205.

Meinzen-Dick, Ruth, and Leticia Nkonya. 2005. Understanding Legal Pluralism in Water Rights: Lessons From Africa and Asia. International Workshop on African Water Laws: Plural Legislative Frameworks for Water Management in Africa, 26–28 Jan 2005.

Mitchell, Bruce. 2002. *Resource and Environmental Management*. Essex: Pearson Education Limited.

Norgaard, Richard B. 2010. Ecosystem Services: From Eye-Opening Metaphor to Complexity Blinder. *Ecological Economics* 69: 1219–1227.

Ostrom, Elinor. 1990. *Governing the Commons: The Evolution of Institutions for Collective Action*. New York: Cambridge University Press.

Ostrom, Elinor. 2003. How Types of Goods and Property Rights Jointly Affect Collective Action. *Journal of Theoretical Politics* 15 (3): 239–270.

Ostrom, Elinor. 2007. A Diagnostic Approach for Going Beyond Panaceas. *Proceedings of the National Academy of Sciences* 104 (39): 15181–15187.

Ostrom, Elinor. 2009. A General Framework for Analyzing Sustainability of Socio-Ecological Systems. *Science* 325: 419–422.

Ostrom, Elinor. 2010a. Beyond Markets and States: Polycentric Governance of Complex Systems. *The American Economic Review* 100 (3): 641–672.

Ostrom, Elinor. 2010b. Polycentric Systems for Coping With Collective Action and Global Environmental Change. *Global Environmental Change* 20 (4): 550–557.

Ostrom, Elinor, and Roy Gardner. 1993. Coping With Asymmetries in the Commons: Self-Governing Irrigation Systems Can Work. *The Journal of Economic Perspectives* 7 (4): 93–112.

Pearce, David. 1998. Auditing the Earth: The Value of the World's Ecosystem Services and Natural Capital. *Environment: Science and Policy for Sustainable Development* 40 (2): 23–28.

Porter, Theodore M. 1995. *Trust in Numbers: The Pursuit of Objectivity in Science and Public Life*. Princeton: Princeton University Press.

Pradhan, R., and R. Meinzen-Dick. 2003. Which Rights Are Right? Water Rights, Culture, and Underlying Values. *Water Nepal* 9 (19) (1/2): 37–61.

Priscoli, Jerome Delli. 1996. The Development of Transnational Regimes for Water Resources Management. In *River Basin Planning and Management*, ed. M.A. Abu-zeid and A.K. Biswas, 19–38. Calcutta: Oxford University Press.

Reuss, Martin. 2003. Is it Time to Resurrect the Harvard Water Program? *Journal of Water Resources Planning and Management* 129 (5): 357–360.

Ringler, Claudia, Anik Bhaduri, and Richard Lawford. 2013. The Nexus Across Water, Energy, Land and Food (WELF): Potential for Improved Resource Use Efficiency? *Current Opinion in Environmental Sustainability* 5 (6): 617–624.

Ringler, Claudia, Dirk Willenbockel, Nicostrato Perez, Mark Rosengrant, Tingju Zhu, and Nathanial Matthews. 2016. Global Linkages Among Energy, Food, and Water: An Economic Assessment. *Journal of Environmental Studies and Sciences* 6 (1): 161–171.

Robertson, Morgan. 2007. Discovering Price in All the Wrong Places: The Work of Commodity Definition and Price Under Neoliberal Environmental Policy. *Antipode* 39 (3): 500–526.

Robertson, Morgan. 2012. Measurement and Alienation: Making a World of Ecosystem Services. *Transactions of the British Institute of Geographers* 37 (3): 386–401.

Russo, Kira Artemis, and Zachary A. Smith. 2013. *What Water is Worth: Overlooked Non-Economic Value in Water Resources*. New York: Palgrave Macmillan.

Sax, Joseph L. 1994. Understanding Transfers: Community Rights and the Privatization of Water. *West-Northwest Journal of Environmental Law and Policy* 1: 13–16.

Schmidt, Jeremy J. 2017. *Water: Abundance, Scarcity, and Security in the Age of Humanity*. New York: New York University Press.

Schmidt, Jeremy J., and Martha Dowsley. 2010. Hunting With Polar Bears: Problems With the Passive Properties of the Commons. *Human Ecology* 38: 377–387.

Schmidt, Jeremy J., and Dan Shrubsole. 2013. Modern Water Ethics: Implications for Shared Governance. *Environmental Values* 22 (3): 359–379.

Schorr, David. 2012. *The Colorado Doctrine: Water Rights, Corporations, and Distributive Justice on the American Frontier.* New Haven: Yale University Press.

Scott, Christopher A., Matthew Kurian, and James L. Wescoat. 2015. The Water-Energy-food Nexus: Enhancing Adaptive Capacity to Complex Global Challenges. In *Governing the Nexus: Water, Soil and Waste Resources Considering Global Change*, ed. Matthew Kurian and Reza Ardakanian, 15–38. Dordrecht: Springer.

Stoler, Justin. 2017. From Curiosity to Commodity: A Review of the Evolution of Sachet Drinking Water in West Africa. *Wiley Interdisciplinary Reviews: Water.* doi:10.1002/wat2.1206.

Trawick, Paul. 2010. Encounters With the Moral Economy of Water: General Principles for Successfully Managing the Commons. In *Water Ethics: Foundational Readings for Students and Professionals*, ed. P.G. Brown and J.J. Schmidt, 155–66. Washington, DC: Island Press.

Trawick, Paul, Mar Ortega Reig, and Guillermo Palau Salvador. 2014. Encounters With the Moral Economy of Water: Convergent Evolution in Valencia. *Wiley Interdisciplinary Reviews: Water* 1 (1): 87–110.

Turpie, J.K., C. Marais, and J.N. Blignaut. 2008. The Working for Water Programme: Evolution of a Payments for Ecosystem Services Mechanism That Addresses Both Poverty and Ecosystem Service Delivery in South Africa. *Ecological Economics* 65 (4): 788–798.

United States. 1936. *Flood Control Act. 74th Congress, Session II, Chapters 651, 688.* Washington, DC: United States Printing Office.

Water, U.N. 2014. *World Water Development Report 2014: Water and Energy.* Paris: UNESCO.

White, Gilbert F. 1978. Resources and Needs: Assessment of the World Water Situation. In *Water Development and Management: Proceedings of the United Nations Water Conference*, vol. 1, ed. A.K. Biswas, 1–46. New York: Pergamon Press.

White, Gilbert F., David J. Bradley, and Anne U. White. 1972. *Drawers of Water: Domestic Water Use in East Africa.* Chicago: The University of Chicago Press.

Wilkinson, Charles F. 1989. Aldo Leopold and Western Water Law: Thinking Perpendicular to the Prior Appropriation Doctrine. *Land and Water Law Review* 24: 1–38.

World Economic Forum. 2011. *Water Security: The Water-Food-Energy-Climate Nexus.* Washington, DC: Island Press.

Zarfl, Christine, Alexander E. Lumsdon, Jürgen Berlekamp, Laura Tydecks, and Klement Tockner. 2015. A Global Boom in Hydropower Dam Construction. *Aquatic Sciences* 77 (1): 161–170.

Zwarteveen, M., and R. Meinzen-Dick. 2001. Gender and Property Rights in the Commons: Examples of Water Rights in South Asia. *Agriculture and Human Values* 18: 11–25.

CHAPTER 4

Societies

Abstract The institutions of global water governance were shaped by social and political conflicts that frequently arose in the context of cultural difference. This chapter is organized according to several aspects of how global water governance has sought to structure social differences. These include: using initial notions of water stress and scarcity to order human-water relations, linking water scarcity to compounding problems of risk and water security, promoting procedural norms regarding transparency and participation as a key to "good" water governance, and encouraging uniformity in values through discourses of water ethics and as the basis for interpreting the implications of the Human Right to Water and Sanitation passed by the UN in 2010. Across these domains, contested social relations regarding gender, class, race, and politics are intrinsic to the identification, articulation, and programs that seek to address global challenges in water governance.

Keywords Social values · Ethics · Scarcity · Security · Good governance
Human right to water · Race · Class · Gender

No list of adjectives can exhaust the ways in which water is important to human societies. It is not only life-giving, but also a well-spring for metaphors that both describe and mediate all kinds of social, political, and economic relationships: from the humility required for leadership in ancient Chinese traditions—where the most powerful leaders, like the most

© The Author(s) 2017   83
J.J. Schmidt and N. Matthews, *Global Challenges in Water Governance*,
Global Challenges in Water Governance, DOI 10.1007/978-3-319-61503-5_4

powerful waters, are those that take the lowest position with respect to others (Allan 1997); to the demands for liquidity in global financial markets (Langley 2016); to from clashes between nonviolent indigenous water protectors with state and private security forces; to chronic mining pollution affecting the livelihoods of entire regions—sometimes for decades (Kirsch 2014); to the river Xanthus that rages against Achilles after his ruthless slaughter of youths in the Iliad. Across these varied domains of materiality and myth, water matters for understandings of subjectivity, social relations, and the symbolic meanings through which determinations of the good life are made. In view of this diversity, anthropologists have suggested treating water as a "total social fact" in recognition that decisions in one domain, such as governance, sustainability, or management, often have implications for others, such as health, religion, economics, or custom (Orlove and Caton 2010). These connections are often unanticipated, even surprising, yet reflect the many facets through which social relations to water intersect with considerations of gender, class, and race. Further, these connections are influenced by social and political structures that crisscross over a range of sectors and scales, from health and education to resource extraction, international relations, and municipal service provision (to name only a few) (see also Gandy 2015; Wagner 2013; Kaika 2005; Mehta 2010; Blatter et al. 2001). In view of this wondrous and at times wicked complexity, this chapter considers how global water governance has sought to mediate water's social relationships.

When it comes to human societies, global water governance has focused its attention on the structures that affect social and political life rather than on choosing among the many different ends individuals or groups may have with respect to water. In a normative sense, governance focuses less on adjudicating complex cases and instead emphasizes procedural techniques that structure how individuals or groups may navigate complex social or environmental contexts. In part, the reluctance to directly adjudicate among different substantive positions was an outcome of the crisis of legitimacy that many states in the global North experienced in the 1970s when, after decades of economic growth, crises of economics and the environment arose simultaneously (Habermas 1973). Given that the foundational conferences and discussions around international water management took shape in this context, it is perhaps not surprising that later structures of global water governance reflect the generic response to social conflicts that emerged in the past four decades. A central aspect of this response has been to expand the number of actors involved in decisions in order to enhance political legitimacy. This has taken place through increasing the number of economic decision makers through markets, as discussed last chapter, but also through increasing the number of stakeholders involved in setting policies and assessing existing practices.

Of course, crises and conflicts in global governance have a much longer history. In fact, they are intrinsic to the international institutions through which global governance programs arose in the twentieth and twenty-first centuries on both environmental and non-environmental issues (Murphy 1994). To understand contemporary global water governance, it is necessary to acknowledge that conflicts are not limited to the choice or application of rules in different social contexts or to the institutional mechanisms used to make them operational. Conflicts run far deeper: from those rooted in the post-colonial contexts that beset the late twentieth century to those emerging in the twenty-first-century acknowledgment that humans have so massively appropriated the planet's resources that they are radically reshaping the Earth system itself. Some of these differences, as the last chapter showed, involve various kinds of economies. Many, however, do not. Still yet other differences form the moral and political basis for acceptable economic arrangements. As such, this chapter is organized according to several key vectors along which global water governance has sought to structure social differences by emphasizing procedural norms. These include: (1) using initial notions of water stress and scarcity to order governance relations, (2) linking water scarcity to compounding problems of risk and security, (3) advocating for procedural values, such as transparency and participation, as key good water governance, and (4) explicitly seeking to influence issues of water and ethics as well as to interpret the implications of the Human Right to Water and Sanitation passed by the UN in 2010.

As the concluding section of the chapter considers, the procedural emphasis of global water governance has itself raised numerous problems, principally because the procedural aspects of social and political structures can also have exclusionary effects and in fact further marginalize certain individuals or groups. For instance, treating "gender" as a distinct, stand-alone category may further marginalize women and girls by ignoring the gendered—frequently masculine—ways that governance structures affect social life and water access in ways that prevent women from participating in governance exercises (Tortajada 1999). For example, property institutions that exclude or marginalize women may limit access to (or power over) water. In such cases, no matter how much development promotes gender equity the fact remains that real equality demands a structural approach to social change. Thought of in the structural terms that occupy the central concerns of governance, many of the social and political structures that marginalize women have a familial resemblance to other forms of oppression, such as those based on race or class (Gaard 2001). Not infrequently, multiple forms of oppression intersect with one another. This further complicates, if it does not render inadequate, attempts to solve water dilemmas through technocratic solutions or strategies that prioritize institutional design over social and moral obligations. In response, many advocates challenging the current arrangements of global water governance call for attention to water justice and for forums in which substantive values are not evacuated from structural considerations in the name of procedural efficiency.

WATER SCARCITY

On its face, water scarcity seems straightforward: a ratio between limited water resources and the water needed for human sustenance and, ideally, flourishing. But this is about as far as agreements go, if indeed they get this far. To begin with, water resources are limited in numerous ways. There are obvious cases of aridity and seasonal or cyclical droughts, but there are also cases of human-induced limits that cover the spectrum from: the effects of climate change on precipitation patterns and increased demands for evapotranspiration, to land-cover changes altering the timing and quality of water flows, to demands for different forms and scales of manufacturing from the artisan to the industrialist, to social and political inequality affecting *for whom* water is limited. Even if we stop short of exploring differences over concepts of human flourishing, it is clear that different assumptions and calculations of water scarcity produce different accounts—with different implications for governance—by virtue of how they order social relationships.

As the introductory chapter noted, water scarcity has been a perennial feature of global water governance since the UN Conference in Mar del Plata. By electing to treat it in this chapter rather than solely in terms of economies, water scarcity can be seen as a governance proposition rather than simply a supply/demand calculation. This also more accurately reflects how water scarcity was taken up in global governance. Indeed, and perhaps somewhat surprisingly, when water was first declared scarce in 1977, there was no agreed upon metric to measure it. Instead of being a rational conclusion to a set of objective calculations, water scarcity was a judgment made based on overall assessments of what was and wasn't known. In fact, the first metric of water stress to gain widespread consensus did not appear until over a decade later when, in 1989, Malin Falkenmark and her colleagues defined: (1) water stress as existing in cases of less than 1700 m^3/year of renewable water per person, (2) water scarcity as existing in cases of less than 1000 m^3/year, and (3) absolute scarcity as existing in cases below 500 m^3/year of renewable water per person annually (Falkenmark et al. 1989). This early metric of water scarcity remains one of the most influential even though it has been critiqued along several dimensions as new approaches to water scarcity have been offered.

Part of the reason that the Falkenmark indicator retains its salience is that the data for it are concrete and available to policy makers (see for an overview: Rijsberman 2006). As such, it does not suffer the difficulty of incorporating judgments regarding social factors that many other indicators demand. The corollary, of course, is that it also does not fully incorporate these social contexts either, which presents its own limitations. For instance, Leif Ohlsson (2000) argued that a determination of Social Resource Water Stress/Scarcity provided a more comprehensive measure that could be calculated by combining the UNDP Human Development Index with hydrological data. From this perspective, water scarcity calculations must also include the general capacity that societies have for adapting to different circumstances, whether these are environmental, social, or political. Likewise, Sarah Wolfe and David Brooks (2003) argued that water scarcity should be seen in terms of: (A) first-order scarcity based largely on biophysical supplies where policy options were limited largely to engineering solutions, (B) second-order scarcity based on demand side calculations where options for governance could be increased through various economic mechanisms, and (C) third-order scarcity, where multiple routes of economic and social adaptation are pursued within biophysical limits. Additional indices linking water to societies have been developed, such as the Water Poverty Index, which gives explicit attention to the plight of poor households in which access to safe and secure water is especially challenging (Sullivan 2002).

In addition to contests over which metrics best reflect the challenges of water scarcity in social contexts, many scholars have identified how water scarcity is often not a natural fact but, rather, a condition produced through various social and political factors and the effects they have on people and the environment. Water scarcity, in this view, should be understood in reference to broader factors of political economy, such as the water demands required for industrial production, and especially capitalist techniques for valuing—and accumulating value from—water. From this perspective, science, politics, economics, environmental conditions, and the infrastructure needed for production are all factors that combine to produce water scarcity, often in places where metrics would neither predict nor reveal it (Bakker 2004; Swyngedouw 2005). Understanding water scarcity in this way, however, is not only a shift in what is measured but also one of approach. That is, instead of designing, applying, and testing a metric that would measure and describe a place as water scarce or not, political economy approaches use water scarcity to name and evaluate social, political, and environmental processes that have detrimental social or environmental effects. In this case, the inequalities that beset different contexts and the differential power relationships that affect water are critical to giving an account of water scarcity. These inequalities can arise from initial or unequal allocations of water, social norms that exclude or marginalize different groups, or combinations of the two with other factors (i.e., economic, political, or environmental) that render water scarce for certain individuals or groups but perhaps not for societies as a whole.

Despite the contingent factors that affect how and to what extent water scarcity affects societies, the hydrological contexts affecting water allocation and distribution are often deployed to render water scarcity as a natural outcome of supply and demands constraints. This veneer of naturalness, however, frequently belies the fundamental social inequalities of water scarcity, particularly the critical implications of inequality for women and the poor (Mehta 2005, 2010). In this regard, water scarcity often refracts issues of social power that affect water use decisions and the structures through which decisions about limited water resources are made. At community through to global scales, these considerations of power imply that determinations of, and responses to, water scarcity always entail social judgments for measuring and responding to water challenges (Schmidt 2012). For example, Canada is frequently identified as a nation richly endowed with a large share of the world's fresh water resources. Indeed, it boasts thousands of lakes and rivers and a relatively low population density. Yet, indigenous peoples across the country are chronically under boil-water advisories and face water pollution threats from industrial facilities, mining practices (both ongoing and abandoned), and municipal wastewater. In this regard, water scarcity in Canada is unequal and unjust. To understand it also requires understanding the governance structure operating in Canada, which is premised on the dispossession of indigenous peoples from their lands, resources, and systems of self-governance (Phare 2009; Boyd 2011; Desbiens 2013). In Canada, decisions about water scarcity and other natural resources must be understood, therefore, in terms of how governance structures systemically marginalize indigenous peoples. This structure can affect even ambitious attempts to improve governance. For example, the Canadian province of Alberta funds stakeholder governance activities through its provincial strategy known as *Water for Life*. These funds arrive at local watershed councils through financial contracts that stipulate all knowledge produced through governance exercises is the property of the government. However, for indigenous peoples who do not wish for their traditional knowledge to become government property, these contractual arrangements provide clear incentive to decline to participate in these exercises (Matthews and Schmidt 2014). The result is that the structure of Alberta's water governance programs can marginalize indigenous peoples: They either surrender to the government their intellectual property or are rendered voiceless in water governance exercises that proceed without them.

Increasingly, water scarcity is being connected not only to human demands, but also to the connected demands of human, non-humans, and Earth system processes (Jaeger et al. 2013). In fact, Malin Falkenmark and her colleagues had initially called for attention to both water and land issues in the 1980s, yet the relatively narrow views of water scarcity that were made operational in the 1990s frequently did not explicitly connect land use, water, and the environment. In the new millennium, however, it has become increasingly evident that water scarcity problems demand both innovative solutions and careful appraisals of the human and non-human contexts that affect sustainability (Jarvis 2013; Richter 2013). For instance, in 2006, the UNDP (2006) explicitly connected water scarcity to issues of gender equality and the uneven relations of social power that affect access to water. As the complexity of water scarcity calculations multiplied, a suite of studies on the growing human impacts on the global water system also connected human-driven water scarcity to issues of security for people, rivers, and global biodiversity (e.g., Vörösmarty et al. 2010). The resulting problem of water security was twofold: (1) Failure to address water scarcity could lead to social and political conflict, and (2) failure to address water scarcity could also undermine the environmental conditions upon which sustainable societies may develop. The combined concerns over conflict and environmental conditions led to growing appreciation of complexity in water governance. It would no longer be possible to continue status quo practices that impair water systems for initial gains and then later repair them (only) with respect to the consequences that appear most politically important (Vörösmarty et al. 2015). This approach, as many scholars and practitioners increasingly acknowledged, both compounded risks and produced a more basic challenge of water security.

WATER RISK AND SECURITY

When it comes to water security, the first thought of many people is that, in the future, "water wars" may arise under conditions of intense political conflict over water. These ideas circulate widely in popular imaginations, such as in post-apocalyptic movies and novels like *Mad Max: Fury Road* (Warner Bros. Pictures) and *The Water Knife* (Bacigalupi 2015). Indeed, there have been intense conflicts over water. But water has also been a beacon for peace. Rather than adjudicate which way of thinking may be more fruitful—water wars, water for peace, or some more nuanced position on the politics and challenges of transboundary water issues (see Blatter et al. 2001; Mirumachi 2015)—programs for global water governance take a much wider view of water risk and security. As early as 1995, the World Bank (1995) connected water scarcity and water security to the challenge of achieving (and the need for) integrated water resources management in the Middle East and North Africa. By the start of the new millennium, at the World Water Forum in The Hague, water leaders from around the world drew on the history of water management from Mar del Plata, Dublin, and the sustainable development conference in Rio united around "...one common goal: **to provide water security in the 21st Century**" (Ministerial Declaration of the Hague 2000, p. 1).

Since the declaration at The Hague, water security has supplemented, and arguably supplanted, water scarcity as a driving proposition of global water governance discourse. In part, this is because water security escapes some of the thornier demands of water scarcity—such as relying on metrics that, as we saw above, may be subject to challenges owing to the judgments they inevitably entail. Water security is also somewhat more capacious. It allows issues of scarcity to be connected to particular areas of potential conflict (i.e., between sectors or between states) and to particular scales of concern for environmental conditions. Further, the shift in emphasis from scarcity to security also paces the shift from management to governance—managing water scarcity remains a salient concern, but it requires a structural response given the complex ways that water affects environments, economies, and societies. Equally as important, however, is that water security allowed international institutions to maintain commitments to the environmental, economic, and social concerns of sustainability while also navigating the tensions that arose in the early twenty-first century over IWRM (see Chap. 1). In this case, rather than being bound primarily to management programs that focus on the point of decision making, water security turned the focus toward how various social, economic, and environmental systems could be vulnerable to risks that intersected across multiple domains in complex ways (Cook and Bakker 2012). For instance, lack of secure access to water as the result of chemical pollution may have intersecting effects on the livelihoods of fishers, the health of fish, those who consume them, and the social and political systems of governance in which fishers are key constituents. As water security evolved, two camps have emerged regarding how these complex interactions should be understood (Zeitoun et al. 2016). The first seeks to achieve "security through certainty" by quantifying different forms of risks in ways that reduce complexity into comparable criteria for decision making. The second seeks "security through pluralism" by focusing on understanding the intersecting social, economic, or environmental demands on water, which are not easily translated across contexts or reducible to common indices.

Certainty is often a highly desirable characteristic of global water governance. Decision makers must be able to defend decisions that involve both securing reliable, safe, water for populations and providing security against water threats, such as floods (Grey and Sadoff 2007). Organizing water security toward notions of certainty is often accomplished through calculations of risk. These calculations draw on numerous sources, from hydrodynamic modeling of expected ranges and distributions of climate change effects on rainfall and flooding, to estimates of monsoon variability and global hydrological patterns affected by human pollutants (especially aerosols) in the atmosphere, to surveys of financiers regarding perceptions of risks to investments, such as the World Economic Forum's annual risk reports (Lehner et al. 2006; Wu et al. 2013; Das et al. 2015). From this position, and not entirely dissimilar to theories about notions of the "risk society" in general (i.e., Beck 1992), efforts to align water security and global governance seek to reduce complexity wherever possible in recognition that analyses of risks can never entirely eliminate uncertainty. In an era of climate change, however, choosing which risks to measure and identifying how these risks will be responded to within existing governance institutions cannot avoid making value judgments in order to simplify complex scenarios (Conca 2015b).

By contrast, pluralism is a potentially powerful way to link water security to global water governance due to the many pathways and contexts through which water insecurity may arise. By maintaining an open disposition toward both the surprising origins of challenges, and toward potentially novel solutions, this approach aims to bring a wider suite of resources and tools to bear on water security. Local variability and outlier events, for instance, may be normalized by risk calculations that seek certainty by using readily available data compiled at national scales. In such cases, however, the structure of global water governance is misfit with some of the most salient concerns regarding why programs for water security are needed, such as the sub-national or multinational scales that affect water security. Pluralist perspectives, by contrast, seek an amalgam of various types of data and aim to integrate them into a comprehensive account of water security (Bakker and Morinville 2013). To do so, however, requires a platform through which to organize many different kinds of claims. This is far from a straightforward exercise. For instance, seasonal comparisons of droughts and monsoons in India present very different kinds of challenges for understanding water security across a range of different actors, from pastoralists and farmers to manufacturing plants and burgeoning urban centers (Asthana and Shukla 2014). More generally, oscillating security challenges between dearth and deluge also affects multiple sites and concerns regarding flood control, electricity generation, and water supplies for megacities. Drawing all of these factors into water security discussions that also seek accounts of social and political contexts can create very difficult, complex scenarios where the effects of decisions are often unclear.

Differing approaches to water security have strengths and weaknesses: Approaches emphasizing certainty allow for justification regarding policies for reducing or dealing with risks, but they may construe risks in ways that only partially capture what is at stake for societies that

are diverse, plural, and often unequal. Approaches emphasizing pluralism aim to capture what is fully at stake in water security challenges for
ecosystems, societies, and local environments, yet face significant hur-
dles when drafting divergent kinds of data into frameworks for decision
making. These limitations, however, should not only be thought of at
the point of decision making. Instead, they should be understood with
respect to how different approaches to water security structure organi-
zational design, institutional arrangements, and collaborative practices
through which different societies are expected to provide, evaluate, and
implement evidence-based decisions. In this sense, and regardless of
approach, governing water security requires the regularization of various
kinds of practices on everything from data collection and monitoring to
program design, funding, and contracts for infrastructure construction
and maintenance. In short, governing water security crisscrosses numerous elements of human societies, and this has led to an increased interest
on how social norms and procedures function.

PROCEDURAL NORMS, TRANSPARENCY, AND PARTICIPATION

One of the most striking differences between water management and
water governance is in how legitimacy for decisions is gained. For much
of the twentieth century, state-led water management practices gained
legitimacy through a combination of state authority and claims to rep-
resent the public good. Two approaches featured most prominently. The
first was a form of bureaucratic utilitarianism, where the state sought to
coordinate, plan, and implement management programs to maximize
welfare (e.g., cost-benefit analysis). The second was a form of individual
utilitarianism, where the state stepped back from direct intervention in
order to allow individuals more freedom and flexibility in resource allocation and distribution based largely on the assumption that enhanced
individual welfare would lead to overall net social benefits (e.g.,
enhanced use of market instruments) (Feldman 1995; Norton 2005).
These two approaches to water management resonated with approaches
to natural resources management (e.g., forestry) more broadly. By contrast, legitimacy in water governance shifts considerations from the point
of decision making—states versus individuals—to the structures and
processes through which various competing perspectives, demands, and
needs are organized. A key upshot of this shift is that, in water policy as
in other domains, such as finance and social welfare, many states opted

to "set the rules" for decision making rather than to adjudicate among the competing interests that different individuals or groups may have (Krippner 2012; Brown 2015).

One way to describe the shift from management programs that adjudicate among competing goods to water governance programs focused on setting the rules for allocating goods is a "procedural turn" (Schmidt 2014). In the procedural turn, states no longer weigh different substantive values against one another in decisions about the allocation or distribution of water. Instead, states create structures for decision making, such as stakeholder boards or watershed planning councils, in which legitimacy is generated through the procedures used to arrive at decisions. Here, the expectation is that, by having fair procedures for arriving at decisions, participants in water governance exercises will accept the outcomes as legitimate (Schmidt and Shrubsole 2013). Throughout the 1990s and into the new millennium, social psychologists and water practitioners increasingly studied notions of fairness and ways of incorporating shared values into decision-making structures that included a wide range of stakeholders: farmers, city officials, health practitioners, corporations, environmentalists, policy makers, citizens, and more (see Syme et al. 1999; Syme et al. 2008). Increasingly, the goal of governance was to reach consensus across a range of stakeholder positions through processes that encouraged collaboration in the design, implementation, and evaluation of different programs for sharing water across sectors (Sabatier et al. 2005).

The shift toward collaborative and shared governance was not straightforward or uniform. Yet, it, or something like it, was increasingly touted in global water governance programs where notions of "good governance" sought to combine considerations of fairness and consensus with enhanced participation and mechanisms to ensure transparency. These aims fit largely with previous programs of IWRM that, since Dublin, had also pushed for greater participation in water management (Priscoli 2004). Given the difficulties with IWRM at the turn of the new millennium, adaptive management became a favored route to enhancing participation in ways that also fit with understandings of complexity. Adaptive management takes an experimental view toward governance with the aim of enhancing the capacity of social and ecological systems to deal with surprise events or anticipated disturbances (Galaz 2007). By enhancing this capacity, also known as resilience, adaptive management was increasingly seen as a way to breathe new life into the institutional architecture of IWRM by providing a format through which to maintain a holistic disposition while incorporating a more nuanced position regarding the complexity of water challenges.

At the global level, more flexible approaches to water management also fit with an emerging emphasis on "good governance." The focus on good governance emerged alongside the Millennium Development Goals and as the World Economic Forum positioned itself as an independent monitor of the MDGs in the early 2000s (Pigman 2007). As the World Economic Forum and other international networks began to develop principles for good governance, it became clear that regularized procedures for governance not only extended to matters of reaching consensus on water use decisions, but also to fiscal concerns over transparency in the allocation and use of water finance. For example, in 2003, the Global Water Partnership argued that although governance was "intensely political" that, nevertheless, IWRM required new formats for governance that may significantly change "...existing interactions between politics, laws, regulations, institutions, civil society, and the consumer-voter" (Rogers and Hall 2003, p. 5). The result was that during the first decade of the twenty-first century, approaches to complexity aligned procedural techniques for participation, consensus building, and transparency with experimental techniques, such as adaptive management, that sought new procedures through which to enhance the resilience of social and ecological systems.

In this context, international networks of water practitioners and academics began fashioning a template for global water governance. And, while they were careful not to envision global water governance as an external mold to which all societies must conform, this emerging vision nevertheless required—indeed was premised on—reorganizing the dynamics of human societies with respect to water. This set up an inevitable contest between competing visions of social order as different views of what constituted fair procedures came into conflict. For instance, abstract ideals of consensus work best under conditions of equality since they assume that individuals or groups will have comparable resources for advocating their positions as they negotiate in water governance exercises. But across the Global South many argued that the sustainable development paradigm, to which IWRM was so central, had not delivered on many of its key promises and as such significant inequality remained (Mehta and Movik 2014). In fact, the failures of sustainable development to enhance the prospects of many communities had led to new forms of international solidarity that arose in response to threats to water from mining and other activities (Kirsch 2014). The upshot was that the ideal conditions presumed by forms of governance premised on political liberalism—where consensus is achieved through the force of the better argument and communicative forms of reasoning (Habermas 1996)—did not reflect the uneven social realities that existed in many contexts.

In response to the diversity of social orders affecting water, and given the unlikely prospect that any single template would adequately capture this diversity, global water governance shifted once more in its search for legitimacy. This time, it sought to cultivate what are known as polycentric forums for governance. The idea behind polycentric governance is that there is no need for a single central body, or single notion of social order, to ensure legitimacy. Rather, multiple centers of legitimacy may exist and affect the ways in which water courses through homes, neighborhoods, or cities. From the perspective of polycentric governance, the goal is to coordinate multiple formats for generating legitimacy—spiritual, political, economic, historical—in ways that respect the multiple procedures human societies may evolve to share water and other commonly held resources. Perhaps most notably developed by the late Elinor Ostrom, polycentric forms of governance also proposed a way beyond the two prevailing options for governance: states versus markets (Ostrom 2010a, b). In complex social and ecological systems, these two options do not capture the many ways in which water circulates within and across social activities or environments. Thus, global governance programs that aim to enhance participation and which advocate for fiscal transparency are in certain senses more of the same: The former is oriented to solving issues of state-led management and the latter, economics.

At the global scale, the idea that there are many social structures through which legitimacy in governance is produced, and hence many centers through which to pursue improved water governance, was also touted as a way to engage with IWRM without reproducing its shortcomings—in short, because by this time IWRM was entrenched as a key framework for decision making around the world, it could not easily be abandoned without leaving a significant vacuum. A polycentric approach offered a way to connect the appreciation of social and ecological complexity with institutional formats that could potentially coordinate many different procedural norms through which individuals and groups navigated that complexity (Galaz 2007). In this case, rather than require strict procedural norms through which IWRM programs would be connected to new programs of global water governance, polycentric techniques would expand global water governance to incorporate the fact that there are many different social and political structures through which norms are produced and accepted. There are many different ways to produce regular and transparent rules for sharing water, for instance, that do not depend on fiscal accounting or quarterly audits but instead rely on other social procedures. However, the push to expand the canvas of values for good governance from participation, consensus, and transparency to a broader set of values required making explicit the values often held implicitly. This led the international water community to discuss ethics.

WATER ETHICS AND WATER RIGHTS

In 2004, UNESCO published a series of essays on water and ethics. Ostensibly the outcome of a process that had begun in 1997, and published its initial findings in 2000, the UNESCO series set about to establish a shared normative framework for water governance that could accommodate multiple different spheres of values and multiple social structures through which procedural norms generate legitimate (i.e., accepted) governance outcomes (Selborne 2000). Perhaps to maintain consistency with earlier efforts, the UNESCO series largely mirrored the standard categories of sustainable development and IWRM programs, with essays dedicated to the ethical issues regarding water and gender, agriculture, institutions, disasters, ecology, and health (see Priscoli et al. 2004). Regardless of its intent, however, the UNESCO water and ethics series also invigorated a broad response as many scholars and organizations began to make the values and norms that were often implicit in water management and governance an explicit element of discussions and studies of water (Brown and Schmidt 2010). These often focused on the specific considerations of equity that arise locally and as the result of place-based norms evolving over generations as different practices of sharing water are tested and tried (Whiteley et al. 2008). Of course, customary water practices do not always equate to ethical water practices, with the upshot being that the turn to ethics and water governance also brought in new challenges. In Nepal, for instance, urban water development has evolved over time in ways that have exclusionary effects as water issues intersect with issues of class, religion, ecology, and state relationships to civil society (Rademacher 2011).

One of the key challenges that arose in discussions of water ethics was how to incorporate an explicit focus on values with policies and practices that often treated substantive values only implicitly and which are much more comfortable constraining governance discussions to procedural concerns (Groenfeldt and Schmidt 2013). Here, the first task was to show that no system of water management and no structure of water governance are value neutral. As shown earlier in this book, and earlier in this chapter, there are social judgments made at numerous points on issues of cost-benefit analysis, water scarcity, and so on. Once these social judgments are acknowledged, the next step is to recognize that many of them have moral implications. That is, they are connected to moral harms and goods because they affect the water available to individuals and groups and so affect the conditions for biological, social, and political life. For many, moral consideration do not stop with humans, but also extend to non-human species and to healthy ecosystems as well. Sandra Postel and Brian Richter (2003), for instance, have argued that it is vital that water be treated ethically for both humans and nature. Similarly, advocates of polycentric and adaptive forms of water governance, such as Malin Falkenmark and Carl Folke (2010), have argued that ethical considerations are central to understanding how to best manage social and ecological systems that have coevolved together because the resilience of the entire water system depends on thinking about a shared form of ecological and social solidarity. This "hydro-solidarity" is most evident at the scale of the watershed, but must also take into account the multiple scales of social and ecological action that affect local water supply and demand (Falkenmark and Folke 2002).

The place-based and coevolutionary views of ethical norms in water governance have coincided with a resurgence of interest in the unique social histories of water. As others have shown, one of the key constraints on global water governance was that in the 1970s it rejected these types of concerns in favor of attempts to tell universal histories of "water and man" (see Schmidt 2017). Now, after 40 years, the need to rethink the social histories of water has led to careful consideration of different place-based norms, such as the "duty of water" that historically linked water to energy and irrigation (Wescoat 2013a, b). Many similar recoveries of water values have taken place, often linking water and ecology to the many religious norms that have historically figured largely in social order. Historical accounts of the relationships between water, engineering, and social order have been important elements of understanding how particular social, economic, and environmental relationships have arisen and persisted in France (Pritchard 2011, 2012), Germany (Blackbourn 2006), and China (Pietz 2015), to name only a few of the countries that projects of water history increasingly connect to contemporary challenges.

Importantly, the reconsideration of water histories is not a purely academic exercise and is, in fact, often taken up by particular communities themselves. The 2015 papal encyclical Laudato Si', for instance, is a recent declaration that follows numerous recoveries of religious and spiritual traditions for valuing water across many faiths and practices (Pope Francis 2015; Peppard 2014; Chamberlain 2008; Shaw and Francis 2008). Traditions of valuing water, however, often run into what are now entrenched social and political structures for water governance. As a result, the explicit incorporation of new, substantive values to water presents an area of social and political conflict. One of the most trenchant areas of dispute between procedural and substantive values in recent years has followed in the wake of the United Nations Declaration on the Human Right to Water and Sanitation in 2010 (Sultana and Loftus 2012).

For many, the Human Right to Water and Sanitation codifies in international law what are widely recognized intrinsic values that cannot be abrogated in the name of procedural norms, no matter how fairly those norms may be structured. That is, the Right to Water demands recognition of substantive values. For example, the Right to Water has been interpreted as creating a foundational value through which citizens may make claims on their states to maintain water as a public good. Often, these claims are grounded in a value distinction such that states cannot properly discharge the Right to Water through economic instruments. Such instruments, it is argued, fail to respect water's fundamentally public nature even if those instruments are designed to be procedurally fair or to subsidize those unable to pay the full costs of water service delivery. These distinctions and battles can be frustrating for many decision makers, who see water pricing as a way to ensure water is conserved and revenues raised for its delivery—which thereby create more sustainable formats for meeting the obligations demanded by the right to water. Compounding these ethical challenges are the environmental constraints that complicate meeting absolute requirements for water rights in an era where climate change may wreak havoc with expected hydrological variability (Eckstein 2010; Westra 2010). Additionally, the unequal social and political contexts to which procedurally fair norms are applied often exacerbate other structural forms of oppression. In Mumbai, for instance, the context of political inequality means that water's public value—and the infrastructure needed to deliver on that value—is central to how individuals and communities make claims on the state to be recognized as citizens (Anand 2017). Thus, privatizing the means of water service delivery (i.e., infrastructure) may entail the exclusion of individuals from the means through which they make demands on the state and, when successful, thereby realize various rights.

CONCLUSION: WATER JUSTICE

The institutions of global water governance have been shaped within and through social contests and, in some cases, conflict. The ostensible goal of governance is to pursue equity, a norm that has been central to sustainable development and sustainability discourses since the 1980s. In the most recent round of sustainability negotiations in 2015, the Sustainable Development Goals emphasized the challenge of inequality as central to addressing environmental, economic, and social concerns. Of course, the SDGs are not only about water, and this has broadened discussions of social practices, norms, and rights to considerations of how issues of water scarcity, security, and both substantive and procedural values fit with other sectors. In this regard, the discussions of the nexus introduced last chapter increasingly figure directly into conceptualizing and understanding the multiple connections that water has with other resource and economic sectors (Weitz et al. 2014). What the nexus has yet to do, however, is to make room for the fact that not all connections among water and other resource or economic sectors are of the same kind. These connections—their conceptualization and representation—also entail value judgments since governance of the nexus depends on a particular political and social structure and that structure is not neutral. Valuing the connections of water to food, energy, or the climate in purely economic terms may not—likely does not—capture the vast array of social or ecological goods and relationships that these connections support over time, space, and coevolving communities. In this regard, the rhetorical impulse that positions these connections as central to governance represents just one social approach to intersectoral challenges. Another approach has been and continues to be advanced by advocates of water justice.

Water justice, like environmental justice and climate justice, is grounded in the fact that the decisions affecting water have direct, indirect, and unequal effects that disproportionately burden women, racial and ethnic minorities, and communities of lower economic standing (Zeitoun et al. 2014). Closely related, the UN Declaration on the Rights of Indigenous Peoples (UNDRIP) in 2007 recognizes the rights that self-governing peoples have to resources in their territories and the need for free, prior, and informed consent for projects that may affect their environments. The UNDRIP provides recognition that indigenous peoples frequently and systematically have suffered social and environmental injustice. Critically for global water governance, this injustice is structural in nature—it issues from social and political orders that discriminate based on gender, race, and class. Ecofeminists have rightly pointed out that structural injustices often share logics of oppression, such as is evident in how the domination of nature in general and water in specific is structurally similar to the domination of women (Warren 1990; Gaard 2001). In this context, it is less surprising, though no less unacceptable that women bear a disproportionate burden with respect to water and that these burdens are often exacerbated by persistent unequal social and environmental conditions (Sultana 2011). In this sense, the challenge facing global water governance is that its discussions of "water and ethics" cannot remain theoretical or be oriented toward providing a philosophical legitimation for particular institutional formats or preferred governance structures (Schmidt and Peppard 2014). Rather, achieving water justice requires a context-specific approach to recognizing and addressing the intersections of social and environmental inequality (Zwarteveen and Boelens 2014; Zeitoun et al. 2014).

Calls for water justice present an opportunity to connect the procedural and substantive values that are central to how global water governance approaches structural challenges. The concerns extend beyond the examples and literature cited here. Indeed, deep structural challenges are to be found at multiple levels of institutional, political, and social order. The post-colonial context that beset twentieth-century programs of international development, for instance, has often used environmental and water issues to reinforce rather than challenge the North–South power relationships (Escobar 2012). As global environmental challenges become more acute, there is a legitimate worry that these chronic injustices will be added to and amplified not only by the unequal distribution of environmental harms but, potentially, also by governance solutions that do not confront existing inequalities (Schmidt et al. 2016). For example, growing calls for transitions to a "green economy" may be used to override or ignore previous development commitments or, at worst, to make new demands on nations or regions on the weak side of global power relationships (Conca 2015a). Likewise, the pursuit of low-carbon energy sources frequently invokes new calls for hydropower. Often, these ignore or downplay environmental impacts of dams using broader comparisons to much more damaging forms of energy production (see Chap. 2) to justify massive infrastructure projects that displace and dispossess individuals and communities of their sources of livelihood—whether these are in upstream agricultural lands or on downstream fisheries impacted by changes to the aquatic ecosystem (Matthews 2012).

Critical issues of justice also beset the procedures through which political recognition in environmental governance arrangements is achieved (Povinelli 2011). After decades of promises for an increased voice, fairer terms for economic and social development, and environmental policies that will facilitate sustainability, it is increasingly evident that the terms of "political recognition" often maintain and reinforce existing power dynamics. This takes place, for instance, by only providing for forms of political recognition within the liberal compromise of sustainable development, which excludes many indigenous and/or non-liberal approaches to social, political, or legal orders (see Coulthard 2014). In water governance, for instance, the recognition of certain kinds of rights to water for indigenous groups may come at the expense of broader claims to territorial self-governance. Or these rights may be recognized in principle, but subsequently ignored either by those obliged to discharge them (Wilkinson 2010). There have been notable exceptions, such as the successful push for recognition for some rivers as legal persons (e.g., Strang 2014). These declarations suggest room for rethinking the social values of water and their connection to alternate sources of authority that structure governance systems. Islands of success, however, should not be focused on to the exclusion of the sea of injustice that continues to allow water to be used and abused (Adler et al. 2007). It remains the case, as Sandra Postel (1992) argued, that new values for water are critical to the long-term success of societies in managing and governing their relationships to hydrological systems that can be rendered scarce through both neglect and good intentions based on unsound principles.

CONCLUSION

The trajectory of global water governance through considerations of scarcity, security, human rights, and ethics is not a natural progression. These areas of concern were and remain sites of the contest over what is measured, for whom, and with what implications for how sustainable governance should operate (Vos and Boelens 2014). The goal of global water governance is to draw together multiple social perspectives within frameworks that are capacious enough to respect social difference while not overrunning the limitations of shared water at local, regional, or global scales. This is a goal that, in many ways, reflects the outcomes of social contests rather than the end point of rational decision making. This goal, however, needn't be pursued through ideas of scarcity or security, which are just one set of propositions for understanding relationships of water to societies. Although global water governance is now quite far along in the path of using these propositions as a way to order multiple social relations to water, the whole project of global water governance is only four decades old. As such, opening up new areas of thought and action through moral and legal arguments presents an important route to ensuring that global water governance does not foreclose on the hard-won lessons that many societies have achieved in efforts to use and value water wisely.

REFERENCES

Adler, Rebecca A., Marius Claassen, Linda Godfrey, and Anthony R. Turton. 2007. Water, Mining, and Waste: An Historical and Economic Perspective on Conflict Management in South Africa. *The Economics of Peace and Security Journal* 2 (2): 33–41.

Allan, Sarah. 1997. *The Way of Water and Sprouts of Virtue*. Albany, NY: State University of New York Press.

Anand, Nikhil. 2017. *Hydraulic City: Water and the Infrastructures of Citizenship in Mumbai*. Durham: Duke University Press.

Asthana, Vandana, and A.C. Shukla. 2014. *Water Security in India: Hope, Despair, and the Challenges of Human Development*. New York: Bloomsbury Academic.

Bacigalupi, Paolo. 2015. *The Water Knife*. London: Orbit.

Bakker, Karen. 2004. *An Uncooperative Commodity: Privatizing Water in England and Wales*. New York: Oxford University Press.

Bakker, Karen, and Cynthia Morinville. 2013. The Governance Dimensions of Water Security: A Review. *Philosophical Transactions of the Royal Society A* 371: 1–18.

Beck, Ulrich. 1992. *Risk Society: Towards a New Modernity*. London: Sage.

Blackbourn, David. 2006. *The Conquest of Nature: Water, Landscape, and the Making of Modern Germany*. New York: W.W. Norton.

Blatter, Joachim, and Helen Ingram (eds.). 2001. *Reflections on Water: New Approaches to Transboundary Conflict and Cooperation*. Cambridge: Massachusetts Institute of Technology.

Boyd, David R. 2011. No Taps, No Toilets: First Nations and the Constitutional Right to Water in Canada. *McGill Law Journal* 57 (1): 81–134.

Brown, Wendy. 2015. *Undoing the Demos: Neoliberalism's Stealth Revolution*. New York: Zone Books.

Brown, Peter G., and Jeremy J. Schmidt (eds.). 2010. *Water Ethics: Foundational Readings for Students and Professionals*. Washington, DC: Island Press.

Chamberlain, Gary. 2008. *Troubled Waters: Religion, Ethics, and the Global Water Crisis*. Lanham, MD: Rowman and Littlefield Publishers.

Conca, Ken. 2015a. *An Unfinished Foundation: The United Nations and Global Environmental Governance*. Oxford: Oxford University Press.

Conca, Ken. 2015b. Which Risks Get Managed? Addressing Climate Effects in the Context of Evolving Water-Governance Institutions. *Water Alternatives* 8 (3): 301–316.

Cook, Christina, and Karen Bakker. 2012. Water Security: Debating an Emerging Paradigm. *Global Environmental Change* 22 (1): 94–102.

Coulthard, Glen Sean. 2014. *Red Skin, White Masks: Rejecting the Colonial Politics of Recognition*. Minneapolis: University of Minnesota Press.

Das, Sushant, Sagnik Dey, and S.K. Dash. 2015. Impacts of Aerosols on Dynamics of Indian Summer Monsoon Using a Regional Climate Model. *Climate Dynamics* 44 (5): 1685–1697.

Desbiens, Caroline. 2013. *Water From the North: Territory, Identity, and the Culture of Hydroelectricity in Quebec*. Vancouver: UBC Press.

Eckstein, Gabriel. 2010. Water Scarcity, Conflict, and Security in a Climate Change World: Challenges and Opportunities for International Law and Policy. *Wisconsin International Law Journal* 27 (3): 410–461.

Escobar, Arturo. 2012. *Encountering Development: The Making and Unmaking of the Third World*. Princeton: Princeton University Press.

Falkenmark, Malin, and Carl Folke. 2002. The Ethics of Socio-Ecohydrological Catchment Management: Toward Hydrosolidarity. *Hydrology and Earth System Sciences* 6 (1): 1–10.

Falkenmark, Malin, and Carl Folke. 2010. Ecohydrosolidarity: A New Ethics for Stewardship of Value-Adding Rainfall. In *Water Ethics: Foundational Readings for Students and Professionals*, ed. Peter G. Brown and Jeremy J. Schmidt, 247–264. Washington, DC: Island Press.

Falkenmark, Malin, J. Lundqvist, and C. Widstrand. 1989. Macro-Scale Water Scarcity Requires Micro-Scale Approaches: Aspects of Vulnerability in Semi-Arid Development. *Natural Resources Forum* 13: 258–267.

Feldman, David. 1995. *Water Resources Management: In Search of an Environmental Ethic*. Baltimore: John Hopkins University Press.

Gaard, Greta. 2001. Women, Water, Energy: An Ecofeminist Approach. *Organization & Environment* 14 (2): 157–172.

Galaz, Victor. 2007. Water Governance, Resilience and Global Environmental Change—A Reassessment of Integrated Water Resources Management (IWRM). *Water Science and Technology* 56 (4): 1–9.

Gandy, Matthew. 2015. *The Fabric of Space: Water, Modernity, and the Urban Imagination*. Cambridge: MIT Press.

Grey, David, and Claudia W. Sadoff. 2007. Sink Or Swim? Water Security for Growth and Development. *Water Policy* 9: 545–571.

Groenfeldt, D., and J.J. Schmidt. 2013. Ethics and Water Governance. *Ecology and Society* 18 (1): 14.

Habermas, Jürgen. 1996. *Between Facts and Norms: Contributions to a Discourse Theory of Law and Democracy*. Cambridge, MA: MIT Press.

Habermas, Jürgen. 1973. *Legitimation Crisis*, trans. T. McCarthy. London: Heinemann.

Jaeger, W.K., A.J. Plantinga, H. Chang, K. Dello, G. Grant, D. Hulse, J.J. McDonnell, S. Lancaster, H. Moradkhani, A.T. Morzillo, P. Mote, A. Nolin, M. Santelmann, and J. Wu. 2013. Toward a Formal Definition of Water Scarcity in Natural-Human Systems. *Water Resources Research* 49: 4506–4517.

Jarvis, Todd W. 2013. Water Scarcity: Moving Beyond Indexes to Innovative Institutions. *Groundwater* 51 (5): 663–669.

Kaika, M. 2005. *City of Flows: Modernity, Nature, and the City*. London: Routledge.

Kirsch, Stuart. 2014. *Mining Capitalism: The Relationship between Corporations and Their Critics*. Oakland: University of California Press.

Krippner, Greta R. 2012. *Capitalizing on Crisis: The Political Origins of the Rise of Finance*. Cambridge: Harvard University Press.

Langley, Paul. 2016. *Liquidity Lost: The Governance of the Global Financial Crisis*. Oxford: Oxford University Press.

Lehner, Bernhard, Petra Döll, Joseph Alcamo, T. Henrichs, and F. Kaspar. 2006. Estimating the Impact of Global Change on Flood and Drought Risks in Europe: A Continental, Integrated Analysis. *Climatic Change* 75 (3): 273–299.

Matthews, Nathanial. 2012. Water Grabbing in the Mekong Basin—An Analysis of the Winners and Losers of Thailand's Hydropower Development in Lao PDR. *Water Alternatives* 5 (2): 392–411.

Matthews, Nathanial, and Jeremy J. Schmidt. 2014. False Promises: The Contours, Contexts and Contestation of Good Water Governance in Lao PDR and Alberta, Canada. *International Journal of Water Governance* 2 (2–3): 21–40.

Mehta, Lyla. 2005. *The Politics and Poetics of Water: Naturalising Scarcity in Western India*. New Delhi: Orient Longman Private Limited.

Mehta, Lyla (ed.). 2010. *The Limits to Scarcity: Contesting the Politics of Allocation*. London: Earthscan.

Mehta, Lyla, and Synne Movik. 2014. *Flows and Practices: Integrated Water Resources Management (IWRM) in African Contexts*. Brighton: Institute of Development Studies.

Ministerial Declaration of The Hague. 2000. Water Security in the 21st Century. 3.

Mirumachi, Naho. 2015. *Transboundary Water Politics in the Developing World*. London: Routledge.

Murphy, Craig. 1994. *International Organization and Industrial Change: Global Governance Since 1850*. Cambridge: Polity Press.

Norton, Bryan G. 2005. *Sustainability: A Philosophy for Adaptive Ecosystem Management*. Chicago: University of Chicago Press.

Ohlsson, Leif. 2000. Water Conflicts and Social Resource Scarcity. *Physics and Chemistry of the Earth: Part B* 25: 213–220.

Orlove, Ben, and Steven Caton. 2010. Water Sustainability: Anthropological Approaches and Prospects. *Annual Review of Anthropology* 39: 401–415.

Ostrom, Elinor. 2010a. Beyond Markets and States: Polycentric Governance of Complex Systems. *The American Economic Review* 100 (3): 641–672.

Ostrom, Elinor. 2010b. Polycentric Systems for Coping With Collective Action and Global Environmental Change. *Global Environmental Change* 20 (4): 550–557.

Peppard, Christiana Z. 2014. *Just Water: Theology, Ethics and the Global Water Crisis*. Maryknoll, NY: Orbis Books.

Phare, Merrell-Ann S. 2009. *Denying the Source: The Crisis of First Nations Water Rights*. Surrey: Rocky Mountain Books.

Pietz, David A. 2015. *The Yellow River: The Problem of Water in Modern China*. Cambridge: Harvard University Press.

Pigman, Geoffrey A. 2007. *The World Economic Forum: A Multi-Stakeholder Approach to Global Governance*. London: Routledge.

Pope, Francis. 2015. *Laudato Si': On Care for Our Common Home*. Huntington, IN: Our Sunday Visitor.

Postel, Sandra. 1992. *Last Oasis: Facing Water Scarcity*. New York: W.W. Norton.

Postel, Sandra, and Brian Richter. 2003. *Rivers for Life: Managing Water for People and Nature.* Washington, DC: Island Press.

Povinelli, Elizabeth A. 2011. *Economies of Abandonment: Social Belonging and Endurance in Late Liberalism.* Durham: Duke University Press.

Priscoli, Jerome Delli. 2004. What is Public Participation in Water Resources Management and Why is it Important? *Water International* 29 (2): 221–227.

Priscoli, Jerome Delli, J. Dooge, and R. Llamas. 2004. *Water and Ethics: Overview.* Paris: UNESCO.

Pritchard, Sara B. 2011. *Confluence: The Nature of Technology and the Remaking of the Rhône.* Cambridge: Harvard University Press.

Pritchard, Sara B. 2012. From Hydroimperialism to Hydrocapitalism: 'French' Hydraulics in France, North Africa, and Beyond. *Social Studies of Science* 42 (4): 591–615.

Rademacher, Anne. 2011. *Reigning the River: Urban Ecologies and Political Transformation in Kathmandu.* Durham: Duke University Press.

Richter, Brian, D. Abell, E. Bacha, K. Brauman, S. Calos, A. Cohn, C. Disla, S. O'Brien, D. Hodges, and S. Kaiser. 2013. Tapped Out: How Can Cities Secure Their Water Future? *Water Policy* 15 (3): 335–363.

Rijsberman, Frank R. 2006. Water Scarcity: Fact or Fiction? *Agricultural Water Management* 80: 5–22.

Rogers, Peter, and Alan W. Hall. 2003. Effective Water Governance. TAC Background Papers No. 7, Evander Novum, Sweden.

Sabatier, Paul A., Will Focht, Mark Lubell, Zev Trachtenberg, Arnold Vedlitz, and Marty Matlock (eds.). 2005. *Swimming Upstream: Collaborative Approaches to Watershed Management.* Cambridge, MA: The MIT Press.

Schmidt, Jeremy J. 2012. Scarce or Insecure? The Right to Water and the Ethics of Global Water Governance. In *The Right to Water: Politics, Governance and Social Struggles,* ed. Farhana Sultana and Alex Loftus, 94–109. London: Routledge.

Schmidt, Jeremy J. 2014. Water Management and the Procedural Turn: Norms and Transitions in Alberta. *Water Resources Management* 28 (4): 1127–1141.

Schmidt, Jeremy J. 2017. *Water: Abundance, Scarcity, and Security in the Age of Humanity.* New York: New York University Press.

Schmidt, Jeremy J., and Christiana Z. Peppard. 2014. Water Ethics on a Human Dominated Planet: Rationality, Context and Values in Global Governance. *WIREs Water* 1 (6): 533–547.

Schmidt, Jeremy J., and Dan Shrubsole. 2013. Modern Water Ethics: Implications for Shared Governance. *Environmental Values* 22 (3): 359–379.

Schmidt, Jeremy J., Peter G. Brown, and Christopher J. Orr. 2016. Ethics in the Anthropocene: A Research Agenda. *The Anthropocene Review* 3 (3): 188–200.

Selborne, Lord. 2000. The Ethics of Freshwater: A Survey. Paris: UNESCO.

Shaw, Sylvie, and Andrew Francis (eds.). 2008. *Deep Blue: Critical Reflections on Nature, Religion and Water.* London: Equinox.

Strang, Veronica. 2014. The Taniwha and the Crown: Defending Water Rights in Aotearoa/New Zealand. *WIREs Water* 1: 121–131.

Sullivan, Caroline. 2002. Calculating a Water Poverty Index. *World Development* 30 (7): 1195–1210.

Sultana, Farhana. 2011. Suffering for Water, Suffering From Water: Emotional Geographies of Resource Access, Control and Conflict. *Geoforum* 42: 163–172.

Sultana, Farhana, and Alex Loftus (eds.). 2012. *The Right to Water: Politics, Governance and Social Struggles.* London: Routledge.

Swyngedouw, Erik. 2005. Dispossessing H_2O: The Contested Terrain of Water Privatization. *Capitalism, Nature, Socialism* 16 (1): 1–18.

Syme, G., B. Nancarrow, and J. McCreddin. 1999. Defining the Components of Fairness in the Allocation of Water to Environmental and Human Uses. *Journal of Environmental Management* 57: 51–70.

Syme, G., N. Porter, U. Goeft, and E. Kington. 2008. Integrating Social Well Being Into Assessments of Water Policy: Meeting the Challenge for Decision Makers. *Water Policy* 10: 323–343.

Tortajada, Ceilia. 1999. *Women and Water Management: The Latin American Experience.* Oxford: Oxford University Press.

UNDP [United Nations Development Programme]. 2006. *Beyond Scarcity: Power, Poverty and the Global Water Crisis.* New York: United Nations Development Programme.

Vörösmarty, Charles J., P.B. McIntyre, M.O. Gessner, D. Dudgeon, A. Prusevich, P. Green, S. Glidden, S.E. Bunn, C.A. Sullivan, C. Reidy Liermann, and P.M. Davies. 2010. Global Threats to Human Water Security and River Biodiversity. *Nature* 467: 555–561.

Vörösmarty, Charles J., Michel Meybeck, and Christoper L. Pastore. 2015. Impair-Then-Repair: A Brief History & Global-Scale Hypothesis Regarding Human-Water Interactions in the Anthropocene. *Daedalus* 144 (3): 94–109.

Vos, Jeroen, and Rutgerd Boelens. 2014. Sustainability Standards and the Water Question. *Development and Change* 45 (2): 1–26.

Wagner, John R. 2013. *The Social Life of Water.* New York: Berghahn.

Warren, Karen. 1990. The Power and Promise of Ecological Feminism. *Environmental Ethics* 12 (2): 125–146.

Weitz, Nina, Måns Nilsson, and Marion Davis. 2014. A Nexus Approach to the Post-2015 Agenda: Formulating Integrated Water, Energy, and Food Sdgs. *SAIS Review of International Affairs* 34 (2): 37–50.

Wescoat, James L. 2013a. Reconstructing the Duty of Water: A Study of Emergent Norms in Socio-Hydrology. *Hydrology and Earth System Sciences* 17: 1–10.

Wescoat, James L. 2013b. The 'Duties of Water' With Respect to Planting: Toward an Ethics of Irrigated Landscapes. *Journal of Landscape Architecture* 8 (2): 6–13.

Westra, Laura. 2010. Climate Change and the Human Right to Water. *Journal of Human Rights and the Environment* 1 (2): 161–188.

Whiteley, John M., Helen Ingram, and Richard W. Perry (eds.). 2008. *Water, Place & Equity.* Cambridge, MA: MIT Press.

Wilkinson, Charles F. 2010. Indian Water Rights in Conflict With State Water Rights: The Case of the Pyramid Lake Paiute Tribe in Nevada, US. In *Out of the Mainstream: Water Rights, Politics and Identity*, ed. R. Boelens, D. Getches, and A. Guerva-Gill, 213–222. London: Earthscan.

Wolfe, Sarah, and David B. Brooks. 2003. Water Scarcity: An Alternative View and Its Implications for Policy and Capacity Building. *Natural Resources Forum* 27: 99–107.

World Bank. 1995. *From Scarcity to Security: Averting a Water Crisis in the Middle East and North Africa.* Washington, DC: World Bank.

Wu, Peili, Nikolaos Christidis, and Peter Stott. 2013. Anthropogenic Impact on Earth's Hydrological Cycle. *Nature Climate Change* 3: 807–810.

Zeitoun, Mark, Jeroen Warner, Naho Mirumachi, Nathanial Matthews, Karis McLaughlin, Melvin Woodhouse, Ana Cascão, and Tony J.A. Allan. 2014. Transboundary Water Justice: A Combined Reading of Literature on Critical Transboundary Water Interaction and 'Justice', for Analysis and Diplomacy. *Water Policy* 16 (S2): 174–193.

Zeitoun, Mark, Bruce Lankford, Tobias Krueger, R. Tim Forsyth, A.Y. Carter, R.Taylor Hoekstra, Olli Varis, Frances Cleaver, Rutgerd Boelens, Larry Swatuk, Christopher A. Scott, D. Tickner, Naho Mirumachi, and Nathanial Matthews. 2016. Reductionist and Integrative Research Approaches to Complex Water Security Policy Challenges. *Global Environmental Change* 39: 143–154.

Zwarteveen, Margreet Z., and Rutgerd Boelens. 2014. Defining, Researching and Struggling for Water Justice: Some Conceptual Building Blocks for Research and Action. *Water International* 39 (2): 143–158.

CHAPTER 5

Water Futures

Abstract The concluding chapter of the book positions the future challenges of global water governance with respect to the growing recognition that humans are driving processes of global environmental change. Numerous futures are possible in the Anthropocene—the proposed geological epoch recognizing the collective force of human activity on the functioning of the Earth system. It tracks two potential structural responses for governance. The first is an emphasis on managing transitions of socio-technical systems. The second is an emphasis on transforming the conditions for governance. It provides a common framework for both by forwarding the four pillars of the UN Charter as the basis for governance. To date, only two of those pillars have been focused on in global environmental structures—international agreements between nations and social progress (i.e., economic development). The remaining two, ensuring peace and respecting the dignity of human rights, offer a normative basis for structuring future responses to the challenges of global water governance in an era of rapid social and environmental change.

Keywords Anthropocene · Transition management
Transformation: UN Charter · Earth system · Global environmental change

Until 2030, the Sustainable Development Goals will have a large effect on the design and funding of international environmental governance.

© The Author(s) 2017   111
J.J. Schmidt and N. Matthews, *Global Challenges in Water Governance*,
Global Challenges in Water Governance, DOI 10.1007/978-3-319-61503-5_5

Water has an explicit place in the SDGs, but it is also vital to several goals regarding health, food, and energy. It is also a prerequisite to others, such as providing education to those who presently spend hours securing water each day. These goals also all share a common challenge, namely that the conditions for environmental governance are novel in many respects owing to the scale of human impacts on the Earth system (Biermann 2014). As a consequence of these changing conditions, it is not clear that existing institutions are up to the task of meeting the SDGs. Sobering assessments, such as Ken Conca's (2015) recent intervention, have shown how efforts to center environmental governance on a combination of international agreements between nations and social progress—economic development—within them ultimately rely on only two of the four pillars of the UN system. If existing institutions are to successfully deliver on environmental governance goals, then the other two pillars—ensuring peace and human rights—must be mobilized to both navigate the demands of a complex world and reinvigorate international resolve. The political challenge is immense, especially after decades of institutional inertia in sustainable development focused on the slimmer framework of international agreements and economic development that fit more easily with its initial liberal compromise (Fig. 5.1).

In view of converging environmental, economic, and social pressures, taking a new approach to water governance is now more of a necessity than an option. For instance, the scale and pace of environmental change are now described by Earth system scientists as a "no analogue" situation (Steffen et al. 2004). That is, there are no precedents in the geological record for what is fast becoming the baseline condition for global water governance as anthropogenic demands on the planet push it into a new operating space (Rockström et al. 2009). The encroachment of humanity up to, and beyond, the planetary boundaries that have conditioned social and ecological evolution for the past ten millennia must now be addressed in the design and pursuit of sustainability and development programs (Steffen et al. 2015). Increasingly, the complexity of global governance is used to motivate attention to resilience as a globalized framework in which to understand and measure socio-ecological systems (Carpenter et al. 2001; Young et al. 2006; Duit et al. 2010). Yet, incorporating the findings of Earth system science into policies for global water governance is far from straightforward. As discussed throughout this book, establishing institutions for international water management,

---

Charter of the United Nations

We the peoples of the United Nations determined
- to save succeeding generations from the scourge of war, which twice in our lifetime has brought untold sorrow to mankind, and
- to reaffirm faith in fundamental human rights, in the dignity and worth of the human person, in the equal rights of men and women and of nations large and small, and
- to establish conditions under which justice and respect for the obligations arising from treaties and other sources of international law can be maintained, and
- to promote social progress and better standards of life in larger freedom,

And for these ends
- to practice tolerance and live together in peace with one another as good neighbours, and
- to unite our strength to maintain international peace and security, and
- to ensure, by the acceptance of principles and the institution of methods, that armed force shall not be used, save in the common interest, and
- to employ international machinery for the promotion of the economic and social advancement of all peoples,

Have resolved to combine our efforts to accomplish these aims
Accordingly, our respective Governments, through representatives assembled in the city of San Francisco, who have exhibited their full powers found to be in good and due form, have agreed to the present Charter of the United Nations and do hereby establish an international organization to be known as the United Nations.

Fig. 5.1 The Charter of the United Nations

and then global water governance, was challenging enough when the basic parameters of global hydrology were thought well known.

In this context—if histories prove any guide—the "wicked problems" that arose from the difficulties of managing complex scenarios may soon turn into "super wicked problems." That is, problems where conflicts are not only about different perspectives toward issues believed to have rational solutions but, rather, problems so complex that what needs to be known is yet to be determined (Lazarus 2009; Levin et al. 2012). In the Anthropocene, multiple futures are possible and each involves different surprises, vectors of potential change, and uncertainties (Nicolson and Jinnah 2016). Indeed, the Anthropocene challenges conceptions of whether the Earth is governable or not and, if it is, what sorts of values and decisions will be used to govern it (Lövbrand and Stripple 2010). The Anthropocene also challenges the cultural assumptions and political judgments through which governance is understood and institutions are developed. All of this presents an opportunity to link the Earth system sciences to more open and equitable policies. It does not, however, guarantee this outcome. The power relations at work in how water governance challenges (and futures) are framed exert powerful effects on what are considered to be concrete options for water policies even when more open, participatory methods are used (Cook et al. 2013). As alluded to in Chap. 1, there is a difference between declarative claims that certain criteria—participation, transparency, consensus—are inherently good because they frame desirable ends, and reasoned defenses of why following certain courses of action are right. As the philosopher Benjamin Hale (2016, p. 43) argues, "When we slip into frames language, we make a serious category mistake." This mistake arises because the prescriptive claims of governance (i.e., "Be transparent") do not equate with reasons for acting in anything more than a strategic sense (Hale 2016). Thus, they leave critical ethical issues unaddressed regarding why being transparent or participatory is good. The upshot is that governance can meet the criteria of "good governance" without meeting ethical obligations regarding water.

As Conca (2015) highlights, pursuing an adequate vision of environmental governance in the twenty-first century will require rethinking the structure of global environmental governance with the equality of rights and peaceful existence of those historically oppressed as a central goal—not an as an addendum to, or "natural" outcome of, international agreements or economic development (cf. Escobar 2012). Sustainability is not simply a matter of reframing environmental challenges in terms congruent with global environmental change. A key question now presenting itself to global water governance scholars and practitioners is how to understand and govern the types of changes that contemporary conditions demand. Over the past decade, two positions have started to emerge. One emphasizes the notion of transition. The aim of transition management is to consider how the complexity of social and technical systems of governance might be shifted in ways that more fully capture the complexity of the socio-ecological systems that water supports (Pahl-Wostl et al. 2010). The potential here is that new water "entrepreneurs" may thereby develop novel approaches to challenges that take account of the local conditions and technical capacities (Huitema and Meijerink 2009). This is not incremental policy in a new guise, where improving on an existing system proceeds in a piecemeal fashion. Rather, it is a managed reorientation of the social and political structures of governance that result in new arrangements and relationships within societies and among societies, economies, and environments. A second position is one of transformation. The starting point here is to note that not only has the role of government changed in the processes of decentralization over recent decades but also the new expectations on government from decentralized actors present significant challenges (Kettl 2000). In environmental governance, this has included the emergence of supranational institutions like the Global Water Partnership, the World Bank, and the World Economic Forum, all of which have been active in driving the kinds of market-based policies considered earlier in this book, with consequent changes for the role of both government and civil society in decentralized decision making (Sonnenfeld and Mol 2002). In trying to understand transformative aspects of governance, it is critical to recognize that qualitative changes can come from multiple directions—potentially simultaneously—when the dynamics of complex systems reach critical tipping points (Homer-Dixon et al. 2015).

Transition or Transformation? The Twenty-First-Century Challenge

The challenges posed by human impacts on the global water system are immense: The likelihood of increased frequency and intensity of extreme weather events will in all probability punctuate the now chronic water challenges from sea-level rise already besetting the world's large coastal cities (Richter et al. 2013). But cities—people—frequently face significant budget constraints that make infrastructure maintenance challenging, especially after austerity policies in much of the world were used as forms of social discipline after the 2008 financial crisis (Bear 2015). Resistance to certain forms of economic globalization has been especially virulent—even violent—in the water sector (Olivera 2004; Piper 2014). It is not especially difficult to understand why, given that water challenges often expose deep inequalities of race and class. Such was the case in Flint, Michigan, where lead poisoning disproportionately affected African-American neighborhoods after austerity policies led to unsound and unsafe shifts in municipal water sourcing. How should these intersecting challenges be understood and acted upon? Should transformations be embraced even with their attendant uncertainties or should transitions be managed in attempts to reground global governance institutions that have not given equal priority to all four areas of the UN Charter?

Answers to this question will shape the water futures for local communities and the kind of planetary system they collectively share. These answers will affect the conditions for integrating multiple kinds of knowledge across multiple environments, economies, and societies. Moreover, there is no recourse to responses that appeal to what is "natural" to address the emerging challenges of global water governance. Instead, those who argue for pragmatic and contextual realignments through transition management make qualitatively similar judgments to those seeking transformations of political, social, and economic structures. In both cases, the lack of a "natural" framework for governance owing to anthropogenic impacts on the Earth system has, as something of a silver lining, forced political judgments to be recognized as such by preventing a retreat into technical discourses that assume stable environmental conditions. Recognizing the shared basis in judgments of different approaches to water governance challenges is a potential bright spot amid intractable complexity. This is because those historically oppressed by claims about what is "natural" have long struggled for equality and justice and the potential exists to ensure that western sciences are not used, once more, to marginalize others (Schmidt et al. 2016).

Currently, global water governance responds these challenges by seeking as close a fit as possible between the temporal and spatial scales of environmental change and the structures of decision making. The implicit assumption is that effective governance mirrors the nimble, just-in-time models of service delivery that firms have evolved. There is a certain kind of logic to this approach: After attempts to holistically gather all of the demands on water into the singular framework of IWRM failed, the next candidate for holistic management was the set of water–energy–food–climate connections identified by the nexus. Because these connections emerge from complex and changing processes, effective governance is that which is able to respond at the appropriate temporal and spatial scale with as few encumbrances as possible. The implicit judgments involved in this turn, however, must also be made explicit. This is because, at the same time that the nexus attends to connections among water–energy–food–climate, there are innumerable other connections that do not receive the same priority, such as those among different species, or those of settlements whose connections to reliable food, water, or energy are sporadic, or those at the interface of permafrost, land, and sea ice that are fast rearranging Arctic environments with significant implications for human health and livelihoods. This is not to say that global water governance should not attend to the nexus, but rather to note that governing the nexus is not free from political judgments about which connections are attended to and how the scale and pace of change associated with those particular connections affect governance.

Is the nexus transitional or transformational? The question cannot be dismissed by those who argue that the pragmatics of water service delivery, flood protection, and infrastructure development necessitate that water must first have economic value congruent with the global economy before it is determined how transitions or transformations are pursued. Rather, we must take seriously how multitudes of people, including the world's poorest populations, already think and act creatively and collaboratively in their pursuit of water. These efforts provide evidence that water is already highly valued and that the bases for transitions and transformations extend on a much wider spectrum than global water governance has often acknowledged. In this sense, the liberal compromise of sustainable development, in which liberal economic orders have both compelled and constrained approaches to environmental policy, must be relativized as simply one way in which global governance may proceed (cf. Bernstein 2001, 2002). Although there is nothing inherently wrong with valuing water economically—water is central to all human economies—the economic values recognized in the contemporary global economy are only one kind of social value to which water is central. Over the last half century, the emphasis on environmental agreements between nations and the measurement of development primarily in economic terms has crowded out other values in the UN Charter on peace and human rights. Meeting the challenges posed by the novel conditions now shaping and reshaping the global water context will require approaches that not only recognize multiple connections but which appreciate that there are many *kinds* of connections from which transitional or transformational water governance may proceed.

THE ENDS OF GLOBAL WATER GOVERNANCE

The emphasis of this conclusion on incorporating the full suite of values promoted by the UN Charter can be viewed as either transitional or transformational. The two, of course, do not need to be seen as mutually exclusive even if they have different emphases. Rather, they are tools that are helpful to think with as the novel conditions presented by global environmental change reveal the limitations of approaches based on environmental stability and notions of "natural" balances between humans and nature. When these tools are combined with efforts to recover values of peace and human rights, they also have the salutatory effect of showing that the existing structure of global environmental governance, and global water governance as part of it, has realistic and substantive alternative values upon which to move toward more sustainable futures.

Incorporating the full suite of values found in the UN Charter may well aid transitions and ultimately prove transformative for global water governance. This is because the UN Charter already provides a normative basis for international institutions and because the fact that it has so far been given only partial uptake suggests that fully incorporating the full suite of its core values could provide new, substantive levers for change. Should such an undertaking proceed, it will no doubt face steep challenges in establishing new norms for peace and human rights in environmental governance alongside—and likely counter to, at least in some respects—the ways in which international agreements and economic development consolidated the aims, efforts, and conceptualizations of sustainability. The roots of global water governance in western approaches to environmentalism, the imbrication of international policies of IWRM with the approach to economic globalization that has characterized the past four decades, and the frenetic dynamics of global environmental change present daunting roadblocks. But these are roadblocks to any path forward, so they are not reasons to wince at the hard judgments that will need to be made in order to sure that the space between H_2O and the global water cycle provides a space for the flourishing of life and human livelihoods across multiple environments, economies, and societies.

REFERENCES

Bear, Laura. 2015. *Navigating Austerity: Currents of Debt Along a South Asian River.* Stanford: Stanford University Press.

Bernstein, Steven. 2001. *The Compromise of Liberal Environmentalism.* New York: Columbia University Press.

Bernstein, Steven. 2002. Liberal Environmentalism and Global Environmental Governance. *Global Environmental Politics* 2 (3): 1–16.

Biermann, Frank. 2014. *Earth System Governance: World Politics in the Anthropocene.* Cambridge: MIT Press.

Carpenter, Steve, J. Brian Walker, Marty Anderies, and Nick Abel. 2001. From Metaphor to Measurement: Resilience of What to What? *Ecosystems* 4 (8): 765–781.

Conca, Ken. 2015. *An Unfinished Foundation: The United Nations and Global Environmental Governance.* Oxford: Oxford University Press.

Cook, Brian R., Mike Kesby, Ioan Fazey, and Chris Spray. 2013. The Persistence of 'Normal' Catchment Management Despite the Participatory Turn: Exploring the Power Effects of Competing Frames of Reference. *Social Studies of Science* 43 (5): 754–779.

Duit, Andreas, Victor Galaz, Katarina Eckerberg, and Jonas Ebbesson. 2010. Governance, Complexity, and Resilience. *Global Environmental Change* 20 (3): 363–368.

Escobar, Arturo. 2012. *Encountering Development: The Making and Unmaking of the Third World.* Princeton: Princeton University Press.

Hale, Benjamin. 2016. *The Wild and the Wicked: On Nature and Human Nature.* Cambridge: MIT Press.

Homer-Dixon, Thomas, Brian Walker, Reinette Biggs, Anne-Sophie Crepin, Carl Folke, Eric Lambin, Garry D. Peterson, Johan Rockström, Marten Scheffer, Will Steffen, and Max Troell. 2015. Synchronous Failure: The Emerging Causal Architecture of Global Crisis. *Ecology and Society* 20 (3): 6.

Huitema, Dave, and Sander Meijerink (eds.). 2009. *Water Policy Entrepreneurs: A Research Companion to Water Transitions Around the Globe.* Cheltenham: Edward Elgar.

Kettl, Donald F. 2000. The Transformation of Governance: Globalization, Devolution, and the Role of Government. *Public Administration Review* 60 (6): 488–497.

Lazarus, Richard J. 2009. Super Wicked Problems and Climate Change: Restraining the Present to Liberate the Future. *Cornell Law Review* 94: 1153–1234.

Levin, Kelly, Benjamin Cashore, Steven Bernstein, and Graeme Auld. 2012. Overcoming the Tragedy of Super Wicked Problems: Constraining Our Future Selves to Ameliorate Global Climate Change. *Policy Sciences* 45 (2): 123–152.

Lövbrand, Eva, Johannes Stripple, and Bo Wiman. 2010. Earth System Governmentality: Reflections on Science in the Anthropocene. *Global Environmental Change* 19: 7–13.

Nicholson, Simon, and Sikina Jinnah (eds.). 2016. *New Earth Politics: Essays From the Anthropocene.* Cambridge: MIT Press.

Olivera, Oscar. 2004. *Cochamaba! Water War in Bolivia.* Cambridge, MA: South End Press.

Pahl-Wostl, Claudia, Georg Holtz, Britta Kastens, and Christian Knieper. 2010. Analysing Complex Water Governance Regimes: The Management and Transition Framework. *Environmental Science & Policy* 13: 571–581.

Piper, Karen. 2014. *The Price of Thirst: Global Water Inequality and the Coming Chaos*. Minneapolis: University of Minnesota Press.

Richter, Brian, D. Abell, E. Bacha, K. Brauman, S. Calos, A. Cohn, C. Disla, S. O'Brien, D. Hodges, and S. Kaiser. 2013. Tapped Out: How Can Cities Secure Their Water Future? *Water Policy* 15 (3): 335–363.

Rockström, Johan, W. Steffen, K. Noone, A. Persson, S. Chapin, E. Lambin, T. Lenton, M. Scheffer, C. Folke, H. Schellnjuber, B. Nykvist, C. Wit, T. Hughes, S. Leeuw, H. Rodhe, S. Sorlin, P. Snyder, R. Costanza, U. Svedin, M. Falkenmark, L. Karlberg, R. Corell, V. Fabry, J. Hansen, B. Walker, D. Liverman, K. Richardson, P. Crutzen, and J. Foley. 2009. A Safe Operating Space for Humanity. *Nature* 461: 472–475.

Schmidt, Jeremy J., Peter G. Brown, and Christopher J. Orr. 2016. Ethics in the Anthropocene: A Research Agenda. *The Anthropocene Review* 3 (3): 188–200.

Sonnenfeld, David A., and Arthur P. Mol. 2002. Globalization and the Transformation of Environmental Governance: An Introduction. *American Behavioral Scientist* 45 (9): 1318–1339.

Steffen, W., A. Sanderson, P.D. Tyson, J. Jäger, P. Matson, B. Moore III, F. Oldfield, K. Richardson, J. Schellnhuber, B. Turner II, and R. Wasson. 2004. *Global Change and the Earth System: A Planet Under Pressure*. Berlin: Springer.

Steffen, Will, K. Richardson, J. Rockström, S. Cornell, I. Fetzer, Elena M. Bennett, R. Biggs, S. Carpenter, W. Vries, C. Wit, C. Folke, D. Gerten, J. Heinke, G. Mace, L. Persson, V. Ramanathan, B. Reyers, and S. Sörlin. 2015. Planetary Boundaries: Guiding Human Development on a Changing Planet. *Science* 347 (6223): 1259855.

Young, Oran R., Frans Berkhout, Gilberto Gallopin, Marco A. Janssen, Elinor Ostrom, and van der Sander Leeuw. 2006. The Globalization of Socio-Ecological Systems: An Agenda for Scientific Research. *Global Environmental Change* 16 (3): 304–316.

INDEX

© The Editor(s) (if applicable) and The Author(s) 2017
J. J. Schmidt and N. Matthews, *Global Challenges in Water Governance*,
Global Challenges in Water Governance , DOI 10.1007/978-3-319-61503-5

Printed in the United States
By Bookmasters